They Sailed the Skies

THEY SAILED THE SKIES

U.S. Navy Balloons and the Airship Program

J. GORDON VAETH

Naval Institute Press
Annapolis, Maryland

Naval Institute Press
291 Wood Road
Annapolis, MD 21402

© 2005 by J. Gordon Vaeth
All rights reserved. No part of this book may be reproduced or utilized in any form or by any means, electronic or mechanical, including photocopying and recording, or by any information storage and retrieval system, without permission in writing from the publisher.

Library of Congress Cataloging-in-Publication Data
Vaeth, J. Gordon (Joseph Gordon), 1921–
　They sailed the skies : U.S. Navy balloons and the airship program / J. Gordon Vaeth.
　　p.　cm.
　Includes bibliographical references and index.
　ISBN 1-59114-914-2 (alk. paper)
　1. Balloons—United States—History—20th century.　2. Airships—United States—History—20th century.　3. United States. Navy—Aviation—History—20th century. 4. Aeronautics, Military—United States—History—20th century　I. Title.
　UG1373.V34 2005
　359.9'4'0973—dc22

2005018608

Printed in the United States of America on acid-free paper ∞

12 11 10 09 08 07 06 05　　9 8 7 6 5 4 3 2
First printing

To
Thomas Greenhow Williams "Tex" Settle
and
Malcolm Davis Ross

CONTENTS

	Preface	ix
	Acknowledgments	xi
1	1915–1918	1
2	Balloon Racing Resumes	5
3	Lost in Canada	10
4	"Let Go!"	13
5	Enter the Rigids	19
6	Thomas Greenhow Williams Settle	35
7	Another Navy Race	45
8	Lightning and Thunder	50
9	The *Graf Zeppelin*	56
10	1929	61
11	*Akron* and *Macon*	72
12	The Big Win	80
13	Highest Aloft	87

14	*Finis* for the Rigids	95
15	Farewell to Racing	102
16	1935–1941	107
17	World War II	115
18	Postwar	128
	Epilogue	133
	Selected Bibliography	149
	Index	151

PREFACE

During the 1920s and 1930s, the Golden Age of Flight, tens of thousands of people would crowd onto fields and into stadiums to watch the start of a balloon race.

Each year—coinciding with the thunderstorm season—there was an international meet called "the big bag classic," or the James Gordon Bennett International Balloon Race. To decide who would represent the United States in this race, a national race was held some weeks before. Balloonists from the U.S. Navy's lighter-than-air service competed in both races, winning and losing, and writing headlines in the sky.

This is their story, told against the backdrop of the naval airship program, of which they were a part.

An epilogue recounts the post–World War II use of Navy-manned balloons for stratospheric research.

ACKNOWLEDGMENTS

Prime source for the information in this book has been Thomas G. W. "Tex" Settle who, during the 1920s and 1930s, was the leading balloonist in the U.S. Navy.

Following his death in April 1980 at age eighty-five, his wife, Fay, entrusted me with his papers, saying she hoped I would one day write a book about his lighter-than-air experiences. I told her I would. *They Sailed the Skies* is a belated fulfillment of that pledge.

Tex's contributions to this work amounted to more than his letters, flight reports, scrap books, and photographs. Often he and I—one a retired vice admiral and the other a retired Naval Reserve lieutenant—talked about the balloons and airships he flew and the races in which he participated. We were neighbors and close friends in Washington, D.C.

During the immediate post–World War II period, I was fortunate to meet and know some of Europe's most renowned balloonists. They included Belgium's formidable Ernest DeMuyter, France's unforgettable Charles Dollfus, and Britain's great lighter-than-air advocate, Lord Ventry. From them I heard and learned much about "vertical sailing."

My most useful published sources have been Ward T. Van Orman's autobiographical *Wizard of the Winds* and James R. Shock's trilogy, *U.S. Navy Airships 1915–1962; U.S. Army Airships 1908–1942;* and *American Airship Bases and Facilities.*

Illustrations are mostly Navy photos. I am appreciative to the Navy Lakehurst Historical Society for the help given me by its historian, Rick Zitarosa, and by Kevin Pace and Steve Rudowski in providing prints.

Eric R. Brothers of The Lighter-Than-Air Society (Akron, Ohio) rendered a similar favor by searching the society's files and the Goodyear balloon and airship photo archives held by the University of Akron. He was assisted in this by David Wertz and William Cody. John V. Miller, Director of Archival Services for the University, and his associates, Steven J. Paschen, George Hodowanec, and Craig Holbert, played important roles in translating the images located by Brothers into copies for publication.

Photos of the Navy's *Strato-Lab* balloons proved surprisingly difficult to find. Winzen Research, their builder in the 1950s and 1960s, was subsequently acquired by another balloon firm, Raven Industries, and by a packaging materials company, the B.A.G. Corporation. Michael S. Smith of Raven and Julie Janicke Muhsmann of B.A.G. tried earnestly to find me a photo of *Strato-Lab V,* the highest flying of the series (113,740 feet). I want to acknowledge their effort on my behalf.

The historic Strato-Lab flights were made by Mal Ross, Lee Lewis, Charlie Moore, and Vic Prather. Mal's widow, Marjorie, graciously provided me with pictures of Lee and her husband. Harris F. Smith, prospective Navy pilot of *Helios,* the canceled project that started it all, supplied me with one of Moore. Smith also sent a photo of himself with his ballooning companion, William J. "Beaver" Gunther.

Many others—they include Moore, first to fly with a polyethylene balloon and flight scientist on *Strato-Lab IV*—have also contributed information for this book. My appreciation to all!

And, of course, especially to my wife, Corinne, for keeping me on course when bogged down with some ninety years' worth of information and photos.

They Sailed
the Skies

ONE

1915–1918

It was May 1915. Europe was at war. In Germany, zeppelin airships were serving with the Imperial German Navy. In Britain, small nonrigid airships (blimps) were being built to hunt German U-boats.

That month the U.S. Navy contracted for its first airship. The Connecticut Aircraft Company of New Haven built the DN-1, Dirigible, Navy, #1. Navy enthusiasm for its new acquisition was tempered, however, by the fact that no one knew how to fly it. To fill this gap, the Navy bought a balloon, also its first, to use as a lighter-than-air training vehicle. From the Goodyear Tire and Rubber Company of Akron, Ohio, it ordered a free-flying, spherical balloon and contracted with the company to provide flight instruction for two officers.

Lt. Cdr. Frank R. McCrary, prospective commanding officer of the DN-1, was one. Lt. Lewis H. Maxfield, slated to train the crews of future DN-1s, was the other. McCrary qualified first, thereby becoming the Navy's first balloonist.

Goodyear engineers Ralph H. Upson and R. A. D. "Rad" Preston were McCrary's and Maxfield's instructors. They had built Goodyear's first free balloon five years before. In 1913 they had constructed another, which they entered in that year's James Gordon Bennett International Balloon Race in Paris. They won it with a flight that took them across the English Channel to a landing in Scotland.

It wasn't until December 1916 that the DN-1 arrived in crates at its operating base in Pensacola, Florida. After being assembled there, it made its first flight 20 April 1917. McCrary was pilot, and a U.S. Coast Guard officer, qualified in heavier-than-air craft, went along to help out. Neither had flown in an airship before.

DN-1, first U.S. Navy airship. *U.S. Navy Photo*

The DN-1 proved to be so heavy it could hardly get airborne and so prone to leakage that it could scarcely retain the 110,000 cubic feet of hydrogen with which it was filled. It made three flights from its floating hangar and then was wrecked being towed through the water.

A frustrated Navy Department, meanwhile, had been looking at alternate airship designs, including the small, nonrigid, pressure-type airships under development in Great Britain. They were the result of a demand by the First Sea Lord, Admiral "Jackie" Fisher, that the Naval Air Service provide him with an antisubmarine airship . . . and within three weeks! He got it in eighteen days!

The British called these lighter-than-air craft "Sea Scouts." They were more popularly called "blimps." When their taut, gas-filled, bags were flicked by someone's thumb, they responded with a sound that seemed to say "blimp." Quickly acquiring this nickname, nonrigid airships have been called this ever since.

Sea Scouts were simple. They consisted of an airplane fuselage, stripped of its wings and tail assembly, slung beneath a cigar-shaped envelope with a bottom fin attached to it.

Secretary of the Navy Josephus Daniels was impressed by the information received about them from the American Embassy in London. On 4 February

1917, he ordered sixteen similar airships be built. Six weeks later, contracts were awarded. There being no airship industry in America, the work was divided among four companies.

The Goodyear Tire and Rubber Company would build nine airships, the B. F. Goodrich Company five, and the Connecticut Aircraft Company two. The Curtiss Aeroplane and Motor Company would provide the engines, airplane fuselages, and fins. Details varied with so many contractors involved—some ships had single lower fins, some double, for example—but essentially the aircraft consisted of a JN-4 ("Jenny"), lengthened to hold a crew of two or three, and powered by an OX-5 engine.

These airships were 163 feet long, had a useful lift of 1,840 pounds (enough to carry a single Lewis machine gun, but no bombs) and could do forty-seven miles per hour maximum and thirty-five miles per hour cruising. The rubberized cotton envelope that contained their hydrogen measured 84,000 cubic feet. Inside that envelope were two air chambers, called ballonets, totaling 19,250 cubic feet.

Purpose of the ballonets was to maintain a constant gas pressure during flight. When one of these chambers filled with air, it would enlarge and replace the hydrogen that had contracted or otherwise "disappeared" owing to change in temperature or altitude. If the hydrogen expanded, for like reasons, it would press against the ballonet and force air out of it. In this way, by inflating and deflating, air ballonets kept the pressure of the lifting gas within desired limits. Of course there had to be a way to blow air into the chambers. This was done by a scoop in the propeller slipstream that collected it and fed it inside through ducts. To this day, nonrigid or pressure airships (blimps) depend on ballonets to keep their pressure, shape, and rigidity. These internal air systems have changed little over the years, except that the dampers and valves have been improved and electric blowers are used.

The Navy called its sixteen airships the "B-class." The DN-1, whether successful or not, was considered to have been the "A." All sixteen had been delivered by June 1918.

To inflate, rig, and prepare B-ships for flight, Goodyear, at its own expense, built a station outside its hometown of Akron, Ohio. Goodyear named the station Wingfoot Lake for its company logo. B-ships built by the other contractors were also assembled there. It became such a beehive of lighter-than-air activity that the Navy requisitioned it and named it the Naval Air Station, Akron. It would remain such until 1921. The Wingfoot Lake hangar and base remain operational today as the focal point of Goodyear commercial airship advertising.

Not surprisingly, the Navy also contracted with Goodyear for it to train pilots for the B-ships. The first class of eighteen students graduated at Wingfoot

B-type airships. *Author's Collection*

Lake on 17 September 1917 with Maxfield in charge. Seven of the class were ordered immediately overseas to fly French airships on patrols from Paimboeuf, near the Bay of Biscay. No American-built airships saw overseas duty during the war. U.S. Navy airship pilots did, not only in France but also in England.

Most of the regular and reserve force officer pilots who were trained in blimps during the war were assigned to seven airship bases established by the Navy along the eastern seaboard. They were located at Chatham, Massachusetts; Montauk Point and Rockaway Beach, New York; Cape May, New Jersey; Hampton Roads (Norfolk), Virginia; and Pensacola and Key West, Florida. Each airship base had a steel-structured hangar 250 feet long. Situated as they were along waterfronts, they also possessed ramps and hangars for servicing seaplanes.

The Navy's B-ships not only flew antisubmarine missions from these locations, they also flew training missions to qualify officers for their wings and the designator Naval Aviator (Lighter-than-Air). (This classification was later changed to Naval Aviator [Airship].) Pensacola was particularly active as a training center.

During the war years, Navy blimps spent 13,600 hours on patrol. They sank no German submarines, but neither did other naval aircraft operating off the American coast.

TWO

Balloon Racing Resumes

When World War I ended, the Navy and the Army each had balloonists on their rosters. The Navy Lighter-than-Air pilots had first been trained in balloons before qualifying in airships. The Army had used tethered observation balloons over the trenches in France to report enemy activity and spot and direct artillery fire.

The Goodyear and Goodrich companies had produced hundreds of these tethered observation balloons, also known as "kite balloons," awkward-looking and equally awkward-behaving, hydrogen-inflated, sausage-shaped affairs. James R. Shock's definitive history of Army lighter-than-air flight, *U.S. Army Airships, 1908–1942*, tallies 446 officers and 6,365 enlisted personnel as having served with Army observation balloons in France. He adds that 130 personnel were forced to parachute from their baskets when attacked. Sixty-one balloons were set afire and destroyed by German aircraft and artillery. Only one balloonist was killed by enemy action. (The Marine Corps also had observation uses for kite balloons, as did the Navy, which flew them from the sterns of ships.)

Army kite balloonists were also trained in free ballooning. It was training that came in handy when a "kite" broke loose and began drifting away. Thus the Army had a large number of qualified balloonists on hand to compete in balloon racing when peacetime returned. Veterans of the Army's World War I observation balloons called themselves the Bag Vets.

Prior to the war, ballooning had been avidly followed as a sport in Europe. Balloonists of various nationalities competed to see who could fly, not the fastest, but the farthest. Large crowds would gather to watch the takeoffs, particularly in the Gordon Bennett race, which was held annually in the country of the prior year's winner.

International balloon racing had been suspended "for the duration" at the outbreak of the war. Now that peace had returned, Navy and Army balloonists waited impatiently for it to be resumed. Partly they were motivated by the fact that neither service had ever participated in a national or international ballooning competition. In the absence of such, they decided to hold one of their own. It would be a "for-fun" event in which there would be no declared winner.

It took place in September 1919 at St. Louis. Four balloons—three Army and one Navy—entered the race. First to land was an Army balloon, its bag leaking gas like mad. A second Army entry succeeded in reaching Pittsfield, Wisconsin. The third flew a respectable 500 miles, only to land in Lake Michigan, where its two occupants were rescued 20 miles from shore. The lone Navy balloon was from Pensacola and covered 484 miles from St. Louis to its landing in Wisconsin. In keeping with the prerace agreement, no one was pronounced winner.

One month later saw the first postwar national race in the United States. Being a "national," it was open to all pilots, civilian and military alike. It began in St. Louis at the Missouri Air Reserve Field at Grant and Merrimac Streets. Inflation was with coal gas. Because of its lesser lift, the bags were the biggest available: eighty thousand cubic feet.

There was one Navy balloon. To assist its pilot, Lt. H. W. Hoyt, the Navy assigned a meteorologist to accompany him as aide. Francis W. Reichelderfer was one of the few Navy meteorologists still on duty following the war. Earlier he had been based in Lisbon to forecast weather for the Navy's NC flying boats in their flight attempt across the Atlantic. After the NC-4 had succeeded, he had been ordered to the Naval Air Station, Akron, for a fast course in ballooning so he could take part in the upcoming race. Reichelderfer, a future chief of the U.S. Weather Bureau, was once asked why ballooning was important to understanding the weather. "In a free balloon," he replied, "you are effectively a particle of air and one learns a great deal about air currents and especially vertical motion. One develops a feeling for the weather which is difficult to describe and which cannot be obtained by other means."

Goodyear's Ralph Upson and Ward Van Orman were the winners with a twenty-seven-hour flight to Stanbridge East, near Montreal. For Van Orman, who would become America's champion balloonist, it was his first race.

The competition was marked by tragedy. Army Capt. Carl W. Dammann and Lt. Edward J. Verheyden, flying a balloon sponsored by the city of Wichita, were forced down into Lake Huron and drowned.

Another team descended east of Barry Bay, Ontario, and wandered in the wilderness for three days before reaching safety.

The Navy's balloon, despite the meteorological wisdom on board, failed to win, place, or show.

Another national race followed in 1920 at Birmingham, Alabama. Ten balloons took part. The single Navy entry carried Lt. Rolfe Emerson and Lt. Frank Sloman. And what a flight they had!

Over Indiana, in heavy rain and with their sand ballast exhausted, the two tried every way they could think of to stay in the air. Climbing up into the rigging, they undid the ropes that held the basket and let it fall away. Then, crouching on the wooden load ring from which the basket had been hanging, they hung on for dear life, grasping the net that was draped over the bag. They were flying in this mode when they saw the shoreline of Lake Erie approaching.

Knowing they could never make it across, they reached for the valve cord, pulled it, and released gas to descend. When they were just above the ground, they pulled another cord, one dyed red, which was the rip cord. Pulling it tore a panel out of the top of the envelope, emptying it of its gas all at once, and dumping them onto the ground.

Despite all this, Emerson and Sloman didn't win. They were bettered by Lt. Richard Thompson flying Army #1, civilian balloonist Homer E. Honeywell, and Ralph Upson in—guess what?—a Goodyear balloon.

The running of the 1920 "nationals" took a back seat to the great ballooning event of the year: the resumption of the Gordon Bennett international race. For balloonists worldwide, it was "the big one," the race that was often called "the classic."

Millionaire playboy James Gordon Bennett had succeeded his father as owner and editor of the *New York Herald*. An imaginative newspaperman, it was he who, in 1869, sent Henry M. Stanley to Africa to find Dr. Livingstone. He also funded George Washington De Long's ill-fated expedition to the Arctic on board the *Jeanette*.

Bennett was a sportsman, although—except for racing his yacht—he rarely participated in sports himself. He liked to sponsor sporting events: an automobile contest in 1902, a balloon race in 1906, and an aviation race in 1909, all named for him.

For years Bennett lived the life of an American in Paris. He had settled there, having been ostracized by society in America. Quite inebriated, he had urinated into the fireplace at his fiancée's New Year's Eve party. This singular feat not only ended their engagement, but it also required his involvement in a pistol duel with the lady's brother. Fortunately both men were poor shots and missed. Bennett moved to Paris where he ran his newspaper by cable. There he established a European edition of the *Herald*.

The Fédération Aéronautique Internationale (FAI), representing the world's national aero clubs, eagerly accepted his offer to sponsor and provide a trophy for an international aeronautical competition. Somehow, that august body, headquartered in Paris, failed to grasp that Bennett intended it to be for all aircraft types. Missing the sponsor's point, it made it, instead, a distance competition among balloons. Thus the James Gordon Bennett International Balloon Race had come into being, a fluke that caused its sponsor to establish, three years later, a separate contest for heavier-than-air craft.

The misunderstanding by the FAI failed to dampen its enthusiasm, or that of its member sport-flying clubs, for the first Bennett balloon race, scheduled for Paris on 30 September 1906. The organization went about preparing for it with much Gallic verve, intent on making it the grandest *manifestation d'aérostation* ever.

Sixteen balloons went up at five-minute intervals from the Tuileries Gardens. A quarter-of-a-million people looked on. My father was one.

The entries included one American balloon, the *United States*. It was piloted by Frank Purdy Lahm, an active duty Army officer and graduate of West Point, who was making the flight as a private citizen. Henry B. Hersey accompanied him.

Lahm was no novice in ballooning. He held FAI balloon license #4, issued in 1905. He was not originally entered in the race but his father had been. The elder Lahm, a retired Army officer in his sixties, was aghast that there was no American entry. If no one else would represent his country, he would. He rushed about taking ballooning lessons to qualify. His son, learning of this, talked him out of it and took his place in the basket.

After taking off, Lahm and Hersey drifted slowly westward. Then the wind shifted, taking them toward the English Channel. Lahm, believing the wind would persist, decided to take his chances and ride it across to Britain. They did and landed at Scarborough, having outdistanced the best of balloonists from Belgium, France, Germany, Great Britain, Italy, and Spain.

The United States had won the first Gordon Bennett. The trophy was a large silver cup. But America's win put the Aero Club of America (ACA) in a bind. Just recently formed and with only a handful of members, it was suddenly confronted with the challenge of hosting a world-famous international event.

Competitive sports flying was governed by the FAI through its "rules of contest." One of these required participating pilots to possess FAI licenses. The ACA had hardly any such licensed balloon pilots on board. Not even Hersey was licensed (the FAI's rule did not apply to aides). Hersey and other club members quickly undertook to get themselves licenses.

Eventually the ACA—today the National Aeronautic Association (NAA)—was able to get its act together. The 1907 Gordon Bennett was held in St. Louis. Ten balloons competed. It was won by a German who flew 872 miles before landing at Bradley Beach, New Jersey. Any farther and he would have ended up in the Atlantic Ocean!

The German victory moved the next race to Berlin where, in 1908, it was won by a Swiss. Eight nations entered twenty-three balloons, the largest number ever to compete in a Gordon Bennett race.

An American named Mix, in a flight from Zurich, won the trophy the following year (1909).

Nineteen-ten brought another U.S. triumph. Pilot Alan R. Hawley, accompanied by colorful, goateed Augustus Post, rode the winds from St. Louis to Lake Tachotogama, Quebec, a distance of 1,172 miles. Landing after forty-six hours, they wandered five days in the Canadian wilderness before coming upon some trappers who led them to civilization. They had been given up for dead.

Germany won again in 1911. And the next year, France. In the final meet before the outbreak of war, Ralph Upson and "Rad" Preston were the winners, riding from Paris to Bremington, Yorkshire, England.

There was no international balloon racing 1914 through 1919.

When the Gordon Bennett was resumed in 1920 from a field near the Sloss-Sheffield Steel Plant in Birmingham, Alabama, there were seven balloons, three of them American but none belonging to the Navy or Army. The national race that preceded it that year had been intended, like other "nationals," to be an elimination contest to determine who would represent the United States in the next Gordon Bennett. Three Army and one Navy balloons had raced but none had performed well enough to qualify.

The winner of "the big bag classic," when it was held in October 1920, was Belgian Army Lt. Ernest DeMuyter. He landed in the *Belgica* near Burlington, Vermont, eleven hundred miles from his Birmingham starting point. Upson, who had won in 1913, was fifth.

Navy balloonists would have to wait another year for another chance to capture ballooning's international championship.

THREE

LOST IN CANADA

In the late afternoon of 13 december 1920, navy balloon A5598 rose for a training flight from the naval air station at Rockaway Beach, Long Island.

On board were Lt. Louis A. Kloor, pilot, and Lt. Walter Hinton and Lt. Stephen A. Farrell. Hinton and Farrell were seaplane pilots along for the ride. Hinton had been on board the flying boat NC-4 on its 1919 transatlantic flight. To sustain them, they had a dozen sandwiches, crackers, chocolate bars, water, and two Thermos bottles of coffee.

Carried by thirty-five thousand cubic feet of hydrogen in the spherical bag overhead, they drifted inland and north. By about eight o'clock, they were over Wills, New York.

Kloor valved the balloon down for a better look around. They called out to attract the attention of anyone nearby. A man heard them, came out of his house, saw that their drag rope was stuck in a tree, and shouted to them to tell them where they were. When balloonists called down to persons on the ground to ask, "Where are we?" the answer was often, "You're up there in a balloon," but this gentleman was kind enough to be more specific.

Kloor, meanwhile, was busy freeing the rope, and soon they continued on their way.

The wind slackened. The cold rain that had been falling stopped about nine o'clock the next morning, and the sun came out. Its warmth heated and expanded the hydrogen, giving the balloon more lift and a new lease on life.

They were now over a great green forest covered with a blanket of snow. Complete silence surrounded them. Nowhere was there a sign of a human being or a habitation. A pristine and beautiful sight, but potentially a deadly one.

They kept flying, looking more and more desperately for some help on the ground. Their sand ballast was almost gone. They cut the drag rope into lengths, which they dropped over the side of the basket to rid themselves of weight and help them remain airborne. Thermos bottles, the fabric lining of the basket, and the seats all went overboard.

About two o'clock that afternoon, they heard a dog barking. If there was a dog nearby, there also had to be humans. Now was the time to land!

Kloor yanked the valve cord. There was the hissing of hydrogen being released, followed by the sound of the valve doors slamming shut when he let go of the line. The balloon, heavier from the loss of gas, descended. When it hit the ground, it was dragged along by the wind until stopped by a tree.

They found no dog and they found no rescuers.

They were lost in the forest and its underbrush and slush. For their thirst, they drank from "moose licks," holes of standing stagnant water. They tried eating moss. With their few safety matches, they lit fires for warmth, using rotting pieces of logs for fuel.

Farrell was in the poorest condition of all. His toes were cruelly affected, so much so that he removed his trousers to wrap them around his feet. Everyone's clothing was in shreds.

After fours days of wandering, making their way southeast using a compass from the balloon, they came upon human footsteps in the snow. They followed them for five miles. Then, at the mouth of a creek, they saw a man on the other side. He either didn't see them or pretended not to. One couldn't much blame him for being wary of these odd-looking strangers, one without pants, all wearing what was left of uniforms he had never seen before.

Kloor—he was the youngest of them at twenty-three; Hinton was thirty-two, and Farrell forty-five—ran after the man and stopped him. He proved to be a Cree Indian, a trapper named Thomas Mark. After he assured himself that these newcomers had no evil intentions, were genuinely in need of help, and were not revenue agents, he assisted them as best he could.

A trading post of the Hudson's Bay Company was located not far away at Moose Factory, Ontario, on James Bay. Mark helped one of them to the post and, from there, a party set out to rescue the others. Three exhausted Navy fliers were put to bed in the central building where the Hudson's Bay employees were housed.

Kloor, of course, wanted to send a message to the commanding officer at Rockaway Beach. Indian runners carried it to Mattice, twenty miles away, where there was a telegraph station. His message read in part: "Driven by storm, Monday, 13 Dec, west by north to lower Hudson Bay (lat 51–50, long 81–00). Lost in forest 4 days. Only available means of transportation by dog sled which will take about nine days. Will leave here about 27 Dec."

He did not say how far A5598 had flown: 862 miles.

Mattice offered both telegraph and telephone contact with the outside world. When Kloor and the others were transported there, indeed by dog team, the press was not far behind. Overnight the tiny settlement became the source of one of the biggest news stories of the day. Fifty thousand words of copy would be sent from there.

It was at Mattice that an unfortunate incident took place. In telling about the flight, Hinton mentioned that Farrell had become despondent on the trek, had talked of suicide, and even had suggested that the others cut his throat and eat him to help ensure their survival.

When Farrell heard that Hinton had said this, he accosted him, threw him a punch to the jaw, and followed it with a left hook that sent Hinton sprawling across a table. The onlookers who tried to restrain Farrell could not help but see the exhausted condition he was in. He was on the brink of a breakdown.

A *New York Times* reporter was present. The result was a three-banner headline on the paper's front page: "Airmen safe at Mattice, brawl over reports sent home; Farrell, enraged by 'suicide' talk, knocks Hinton down: Congress likely to order investigation of flight."

Navy Secretary Daniels, when he learned of these goings-on, forbade any public statement by the three, unless first cleared by the Department. He also prohibited their selling the story of their experience. A board of inquiry was convened but there was no Congressional investigation.

From Mattice, the three were taken by dog sled to Cochrane, one hundred miles to the east, a stop on the Canadian National Transcontinental Railway. The weather at Cochrane was about the coldest the Canadian North could generate in December.

Back in the States, Farrell was admitted to the naval hospital in Brooklyn where he was diagnosed as suffering from nervous exhaustion. In time he and Hinton overcame their differences. The Navy would say they had been resolved "in a manly way by the giving and accepting of an apology."

Number A5598 was left where it had come down. There it would remain until 1937 when some of it was recovered and returned to the United States. Today what remains is in my attic.

FOUR

"Let Go!"

In america, as the 1920s unfolded, the command "let go!"—the order given to a ground crew to release a balloon—was increasingly heard in the land.

For U.S. Navy balloonists, however, the decade began tragically.

An all-night training flight took off at the naval air station at Pensacola on 23 March 1921. Five men were on board, with Chief Quartermaster E. W. Wilkinson in charge. His enlisted trainees were R. V. Wyland, E. L. Kershaw, and J. P. Elder, plus W. H. Tressey of the Marine Corps. Wilkinson was an experienced airman who knew well the hazards of flying balloons in the Pensacola area where water abounded.

Two messages were received at the station by pigeon from the balloon. The first reported it drifting slowly northwest over the Gulf of Mexico, twenty miles from St. Andrews Bay. The second advised that all sand ballast had been dropped. They were at one hundred feet and descending.

After that: nothing. The balloon and the men were never found.

Two months passed. On 21 May 1921, the Navy participated in the U.S. "nationals," the elimination race that preceded the Gordon Bennett. It originated, as had the 1920 national and international meets, in Birmingham.

There were nine balloons, including one each from the Army and Navy (piloted by Lt. Cdr. L. J. Roth and Lt. H. E. Halland). Upson, who had left Goodyear, was piloting the *Birmingham Semi-Centennial.* His erstwhile pupil, Ward Van Orman, was flying the *City of Akron,* sponsored by its chamber of commerce. Van Orman's aide was Willard F. Seiberling, son of the president of Goodyear Tire and Rubber.

Winds from Birmingham were weak. None of the entries got very far.

Upson won, no doubt helped by the meteorological knowledge of his aide, C. C. Andrus, a chief forecaster in the Weather Bureau who had been loaned to him for the race. They landed at Stewart, Virginia, 423 miles distant. The *Riverview Club of St. Louis,* with pilot Bernard von Hoffman, took second place with 201 miles. Van Orman and Seiberling were third. Roth and Halland finished last.

The Gordon Bennett followed on 18 September. Belgian Ernest DeMuyter had won the year before, so the race originated in Belgium, in Brussels. There were thirteen balloons in it.

To the spectators watching the proceedings, the balloons looked at first like large mushrooms growing on the field. As they watched, the "mushrooms" grew bigger until they finally assumed the spherical shape of gas-filled balloons.

The process of readying one of these "machines," as some people called them, began with unpacking the envelope from its basket, which had served as its shipping container. Each balloon had its own ground crew of a half-dozen or a dozen men. They took the bag, unfolded it, and gently laid it atop a ground cloth that would protect it from stones and stubble on the field. No one was allowed to walk on the fabric without rubber-soled shoes.

With the balloon empty on the ground, a valve, about three feet in diameter, was inserted in a hole at the very top. To avoid striking a spark that would ignite the gas inside, the valve was made of wood. It was clamped and secured to the rim of the hole. It was a "butterfly" type, with two doors hinged back to back. They opened downward to release the hydrogen or coal gas when the valve cord attached to them was pulled. A kind of bungee cord—not metal springs because they could generate the dreaded spark—would return the doors to the closed position when the cord was released.

After the valve was installed, the crew began filling the envelope. If the gas being used was hydrogen, it came from cylinders, if coal gas, it came from local mains. It entered the bag via a hole in the bottom through a fabric inflation sleeve.

The basket, with its payload, would be carried by a cotton net with a diamond-shaped mesh draped over the bag. As gas whooshed into the envelope—not too fast because that could produce a spark—sandbags were hooked on the spaces of the net to keep the inflating bag from breaking loose and shooting skyward. Professional balloonists carefully watched how newcomers handled sandbags. If the newcomers picked up the bags by the necks, instead of from the bottoms, the professionals considered them inexperienced amateurs.

When the gassing was completed, the bottom of the net with the sandbags hanging from it, reached down to where the basket waited. The net's bottom was attached to a heavy wooden ring, the "load ring." The basket, in turn, hung from it.

The basket, known to old-time aeronauts as "the passenger car," was of wicker and about waist-high. It was reinforced with wooden beams along the sides to withstand the beatings from hand landings, crashing through tree tops, and other incidentals of balloon flying. Wooden slats along its bottom helped it move across terrain without being damaged. Basket size varied. A medium size had a floor area of about four by five feet—enough room for the crew who engaged in the race. The rules permitted only two on board: pilot and aide. There were no women balloon racers.

The sandbags, thirty-five pounds each, which had been lowered down the net while the bag was being inflated, were transferred to the basket. Some were hung on its sides. Some were piled on the floor. Some had their contents emptied into a canvas trough from which the crew could throw sand over the side by the handful. Whole bags of sand were dropped only in emergencies. Normally their contents were paid out gradually, even miserly. The amount of ballast remaining would decide how long the flight would be. Also whether the descent would be controlled or be a crash. Getting rid of weight was the only way to stop a descent.

Piloting a free balloon seemed simple enough. Drop ballast to go up. Valve gas to come down. But there was more to it than that. Flying a gas balloon was an art that had to be learned and practiced. It required a working knowledge of the laws governing the behavior of gasses, also of the principles of aerostatics, the science of lighter-than-air flight. One learned, for example, to check a descent when it was beginning, by dropping a handful or trowel-full of sand before the balloon gathered momentum in its fall.

To know whether a balloon was rising or falling, crews carried an instrument called a variometer. It was a kind of rate-of-climb indicator. It and the aneroid altimeter that was also on board were not designed for the vibration-free flight of a balloon. Crewmen had to tap them in flight to shake loose the pointer and obtain an accurate reading.

Another way to judge rate of ascent and descent was by eyeball. Toilet paper, torn into small pieces and thrown over the side, would float to earth very slowly. (It was also a good indicator of wind direction and speed as it fell.) How the paper behaved, whether it stayed with the balloon, dropped away from it, or "fell up," was a good indication of what the balloon was doing. Later, given the increasing concern over the environment, balloonists would take to blowing bubbles over the sides, using a small hoop they dipped into a jar of soapy solution.

Once the baskets were attached, the crews loaded them. The essential paraphernalia normally included brandy and other necessary restoratives for fatigued aeronauts. They took all sorts of items with them, sometimes even a trumpet to express the ecstasy of ballooning while drifting along with the breeze. After everything and everybody was on board, the balloon was "weighed off." This was when the ground crew briefly let go of it so the pilot could determine whether it would rise or not when released. If it was "light," it would. If "heavy," it would not. If it was found to be heavy, sand would be removed until it was buoyant enough to lift itself off.

After weigh-off the balloon was walked out onto the field for launching. Order of takeoff had been predetermined by drawing lots. To be first was not an advantage. The longer a contestant waited for his turn on the ground, the more he could observe the progress of his competitors already in the air. By watching the direction and speed of their drift, he could learn about the winds he'd experience aloft.

The field used by the Gordon Bennett race of 1921 was near where the Germans had outraged the civilized world by executing nurse Edith Cavell in 1915. One of the first entries to rise from it was a team from Italy.

It did so to the accompaniment of cheers, waves, and blown kisses from friends, relatives, and supporters. Peering aft through the basket ropes at the excited crowd they were leaving behind, they failed to notice that their balloon was slowly sinking back to the ground. And heading right for a clump of trees. When they finally realized what was happening, it was too late. Into the trees they went!

But not for long. Dumped into a heap at the bottom of the basket, they regained their feet and composure, liberated their balloon from the branches, and resumed the flight.

Belgian Ernest DeMuyter, winner of the previous year's race, was, to be expected, the favorite of the Belgian crowd. It expected him and his *Belgica* to win. But a freak incident deprived him of the opportunity.

DeMuyter spoke French. His locally supplied ground crew spoke Flemish. One of its members failed to understand when he called out *"lachez-tous!"* ("Let go, everyone!"). The man continued to hold the basket's rim. The balloon started up. The man tried to back away. He couldn't. His belt had become snagged on a grapnel hanging from the basket. (Grapnels served as a kind of brake. Dragging their prongs across the ground or through trees could slow a balloon while it remained in the air. Grapnels could be formidable. They could tear up fence rails, overturn outhouses, and pull off rooftops.) Balloon and grapnel were pulling him up, shouting for help and wildly waving his legs.

On hearing the terrified man's cries and seeing his head just outside the basket, DeMuyter and aide reached out and were able to pull him on board. They had saved his life but the weight he added to the *Belgica* eliminated any chance it could win. DeMuyter ended in twelfth place.

The winner was R. Ambruster, a Swiss, who flew 504 miles from Brussels to Sambay Isle, Ireland. Upson, in what would be his last Gordon Bennett appearance, was third and Van Orman, eighth. The U.S. Navy and Army, having failed to qualify in the "nationals" that year, were not represented.

In the following year's competition (1922) from Geneva, three American balloons were entered: one Navy, one Army, and one civilian. Their pilots were Lt. W. Reed, Maj. Oscar Westover, and Homer E. Honeywell. (Oscar Westover would become a major general and chief of the U.S. Army Air Corps in 1937. Homer Honeywell never won a Gordon Bennett but in 1911 he outdistanced the pilot who did. Barred from participating because his balloon was varnished silk and not rubberized fabric, he took off in it nonetheless, winning it by thirty miles.) They had qualified in the 1922 national meet in Milwaukee. A helium balloon, flown by Lt. Cdr. Joseph P. Norfleet, was in that Milwaukee contest. It was the first time the nonflammable gas had been used in a man-carrying balloon.

Colorful, mustached, DeMuyter was on hand in Geneva to make up for the failure forced upon him the year before. He succeeded by flying 850 miles to Donitza-Gaiova, Romania.

Lieutenant Reed placed second. DeMuyter went farther but his balloon drifted away after he landed. A winning pilot, so said the rules, had to be in control of his balloon on landing. DeMuyter had been but, when he and his aide got out of the basket, it took off. Should he have been disqualified? The judges said no. A contrary decision would have made the U.S. Navy the winner.

DeMuyter's win moved the Gordon Bennett back to Brussels in 1923. There, on 23 September, the twelfth such contest was held. It would be the most catastrophic day in balloon racing history!

Violent thunderstorms surrounded the field as the takeoffs began in late afternoon, the usual starting time. FAI rules did not permit cancellation due to weather. Grim-faced pilots and aides had no alternative other than to fly.

Honeywell never got off the ground. Caught and twisted by the winds, his balloon, *St. Louis,* was ripped open.

U.S. Army balloon, S6, carrying Lt. Robert Olmstead and Lt. John Shoptaw, was blown into a Belgian balloon, *Ville de Bruxelles,* ripping its netting and forcing it out of the race. The American balloon, apparently undamaged, continued to rise.

Amid a theatric setting of lightning flashes, punctuated by thunderclaps, more balloons followed.

Two would be struck by lightning and set afire during or shortly after launch.

Switzerland's *Génève* fell to earth in flames, killing both occupants.

Another bolt hit Spain's *Polar* at thirty-five hundred feet. It killed one of the crew outright. The other rode the flaming bag down. It had caught fire but only the lower half burned. The upper parachuted up into the net and brought the surviving Spaniard down. He jumped to the ground from about one hundred feet and landed in a swamp, breaking both legs.

Over the Netherlands, the Army's S6 was also hit. Olmstead and Shoptaw were killed.

Other balloons were damaged or wrecked. France's *Savoie,* trying to overfly the weather, was so violently forced to the ground that its occupants climbed into the rigging to avoid injury when the basket hit the earth. Pilot and aide of Spain's *Esperia* were injured when they tangled with a power line while landing. The crew of yet another balloon had to be rescued when they descended into the sea.

Navy balloon A6699 was aloft in this atmospheric maelstrom. The pilot was Lt. John B. Lawrence, with aide meteorologist Lt. Francis W. Reichelderfer. Drifting among the cumulonimbus storm clouds over the Netherlands, they became increasingly concerned that they were drifting toward the open sea. When night fell, they came upon lights in a row on the ground. They knew it was the coastline. Balloon A6699 landed in a turnip field twenty-five hundred feet from the Zuider Zee. It hadn't won but it had avoided the tragic fate of so many of the other entries.

Who did win? Ernest DeMuyter, of course. He had dodged the weather to fly six hundred miles from Belgium to Sweden.

FAI rules permitting a delay but not cancellation of a race due to weather continued unchanged.

FIVE

ENTER THE RIGIDS

The Navy ordered thirty C-type airships from Goodyear and Goodrich before the Armistice that ended World War I. Later this blimp order was reduced to ten. The Navy accepted eight and turned the remainder over to the Army.

C-ships were powered by two 125- or 150-horsepower engines. They could do sixty miles per hour. Cruising range was 1,250 miles. Their envelopes of 172,000 cubic feet of hydrogen enabled them to carry a crew of four in addition to four 270-pound bombs.

The "C" was a major improvement over the "B" and undertook many things the older airship could never do: carrying an airplane for one, attempting to fly the Atlantic for another.

On 12 December 1918, the C-1 (Lt. George Crompton, pilot) carried an Army J-4 aloft and released it.

And in May 1919, Lt. Cdr. Emory W. Coil and crew set out in the C-5 to be "first across" the Atlantic. To stretch its range, its engines were modified to burn hydrogen from its envelope as well as gasoline from its tanks. The ship flew nonstop from Montauk Point to St. John's, Newfoundland. Landing there, it was faced with gale winds. There was no mooring mast—the Navy had none at the time—to help ground handlers keep the C-5 under control. They couldn't do so. With no one on board, it was wrenched out of their hands, to drift out over the Atlantic, an airborne derelict never seen again.

A "C" was the first airship of any kind to fly with helium. It was the C-7, in December 1921, from the air station at Hampton Roads.

The C-5 attempts to fly the Atlantic in 1919 but winds tear it from its ground handlers in Newfoundland. *U.S. Navy, Naval History Center*

A six-foot panel inserted in the envelope of a "C" made it into a "D." To make more room in the car, the D-ship had its fuel tanks mounted on the sides of the bag. An incredible arrangement that made for a maintenance and servicing nightmare! Goodrich and Goodyear built the envelopes for the "Ds." The Naval Aircraft Factory, Philadelphia, manufactured the car. For a while, Lt. (jg) Leroy Grumman was the Naval Aircraft Factory's project engineer on the job.

The first five "Ds" were transferred by the Navy to the Army. The D-6 fell victim on 31 August 1921 to a gasoline fire in the hangar at Rockaway Beach. Two other airships, one of them a "C" and the other the H-1, were consumed, as was the hangar.

Between the "D" and the "H," there had been one "E" and one "F," about the size of a "C." They were built by Goodyear on speculation, about the only thing that can be said about them.

The G-class was a World War I design that was never built. It would have been a super-blimp of four hundred thousand cubic feet, armed with a three-inch cannon to hunt submarines.

As for the H-1, it was a small, fat, roly-poly affair intended to serve as an airship or as a kite balloon. The Navy didn't bother to replace it.

The Navy's postwar aircraft inventory included a dozen or so foreign-built airships, French, British, and Italian, which had been acquired overseas. Few were in condition to fly. The ones that were contributed little to the Navy's lighter-than-air effort.

Meanwhile the months passed: The air station at Chatham was put up for sale. The hangar at Montauk was disassembled and shipped to Cape May where it would be used to enlarge the shed already there. Hampton Roads became a base for airship development. Pensacola prepared to transfer its balloon and airship training to Hampton Roads. Navy blimps no longer operated out of Key West.

Return to civilian life by personnel recruited during the war had, of course, much to do with this retrenchment. But there was something else: the coming of the rigid airships. The hulls of rigid airships, exemplified by Germany's zeppelins, contained an internal framework or skeleton, hence their name "rigid." (Navy men called them "rigids"; the public called them "dirigibles.")

Navy Department interest in lighter-than-air craft was now focused on the extraordinary endurance, range, and load-carrying ability of these giants in the sky. They had demonstrated a scouting usefulness to the Imperial German Navy. They, much more so than the little nonrigids or blimps, appeared to be the lighter-than-air craft of the future. They were aircraft the U.S. Navy wanted to own and operate.

And so did the Army. Brig. Gen. William "Billy" Mitchell had inspected the wreckage of the German naval zeppelin, L-49, that had been forced down intact in France. He was impressed. A plan, believed to be Mitchell's handiwork, was concocted within the War Department to buy a zeppelin from the recently defeated Germany. This was in violation of an earlier decision of the Joint Army and Navy Airship Board that designing and building rigid airships, at least the first ones, was to be a Navy responsibility. Pressure airships—that is, nonrigids or blimps and semirigids (blimps with a keel)—would be the Army's domain.

The scheme involved Lt. Col. William N. Hensley Jr., a senior Army balloonist who had been assigned to be his service's observer on board the British rigid airship R-34 on its return flight from Long Island, New York, to England in July 1919. (A Navy observer, Lt. Cdr. Zachary Lansdowne, had been on board for the westbound Atlantic crossing earlier in the month.)

Hensley's orders were to learn as much as he could about rigids on the R-34 and elsewhere in Europe. He was instructed, furthermore, to purchase the German naval zeppelin, L-72, which the Zeppelin Company, Luftschiffbau-Zeppelin, had completed but never delivered. Its constructors considered it still their property and wanted to sell it.

Hensley was supposed to seal the deal for the 743-foot-long, 2,418,000-cubic-foot-volume "height climber," capable of twenty-thousand-foot altitudes. Knowing what Hensley was up to and not wanting America to possess the newest and best of all airships, the Inter-Allied Aeronautical Commission of Control gave the L-72 to France as a war reparation. How the Germans felt about this was indicated by the graffiti and anti-French obscenities scrawled throughout the L-72 when it was turned over to the former enemy.

Hensley persevered. If he couldn't buy the L-72 from the Zeppelin Company, perhaps he could negotiate with it for a new airship to be built for the Army. In November 1919, he signed a contract with Luftschiffbau-Zeppelin to that effect.

At this point, Secretary of War Newton D. Baker suddenly became aware of what was going on. He ordered the contract cancelled immediately. As yet, there was no peace treaty between Germany and the United States. The two countries were technically still at war with one another. Hensley's contract amounted to dealing with the enemy.

The door wasn't entirely closed to Army participation in rigid airships, however. The Board that had designated the Navy as the cognizant service also directed it to make its rigid airship experience and knowledge available to the Army. The Navy would do so, including giving Army personnel rigid airship flight training in the 1920s.

The 1919 Naval Appropriations Act of Congress provided for two Navy rigid airships and a lighter-than-air station with hangar. Lacking experience in designing and building aircraft of this kind, the Navy looked for one it could buy.

Obtaining one from Germany now seemed out of the question. Britain, on the other hand, had been an experienced builder of rigids since 1911. It had completed a dozen of them. One was being constructed that His Majesty's Government was willing to sell. It was the R-38. The United States bought it for £300,000.

Known to the American Navy as the ZR-2 (Lighter-than-Air, Rigid, #2), the airship was 699 feet long, 85 feet wide, and 2,724,000 cubic feet in volume. Inflation would be with hydrogen, the only suitable lifting gas then available.

The R-38 was not strongly built. Its design ceiling was twenty-two thousand feet. To reach that height, its structure had to be extremely lightweight. High speed maneuvering, especially in dense air, was to be avoided.

On its fourth flight—it was 24 August 1921—the ship was put through turning trials at sixty-three miles per hour (fast for an airship) at twenty-five hundred feet. The rudders were worked hard and at full throw.

The R-38 broke in two. Its forward part fell into the Humber River at Hull, England, its gasoline spilling onto the water and catching fire. Its stern, not on fire, fell slowly, coming to rest on a sand bar.

The fragile, lightweight structure, the maneuvers at high speed and low altitude, and the failure of its builders to calculate aerodynamic loads all contributed to the disaster. In fairness to its designer, C. I. R. Campbell, predicting in-flight loadings was little understood in 1918 when the R-38 was designed. Campbell and his staff, it would appear, simply did not know how to do it. Instead, they based their calculations on static loads, adding what they considered a generous safety factor.

Sixteen Navy men had lost their lives. They included Lt. Cdr. Lewis H. Maxfield, the R-38's prospective commanding officer, and Lt. Cdr. Emory W. Coil, of C-5 distinction. They were part of a special contingent sent from the United States to Howden, where British rigids were based. After accepting the airship, they were to fly it to America where it would be based, at least initially, at Cape May. The reason the blimp hangar at Montauk had been used to enlarge the one at Cape May was to have one big enough to accommodate the ZR-2 when it arrived.

The British loss was no less than the American. It included Air Commo. E. M. Maitland, Leader of Airships, and twenty-three airshipmen.

The R-38, or ZR-2, died before it could be named.

Across the Atlantic, sixty miles south of New York City in New Jersey's pine tree belt, another rigid airship was, meanwhile, under construction to take the ZR-2's place. It was being built at Lakehurst, New Jersey, at the base authorized by Congress to have a hangar and capabilities of handling two rigids at a time. Before becoming a naval air station, the site had been an ordnance test locale and a chemical warfare center.

A steel hangar, covered with asbestos panels to minimize temperature changes inside, had been built there. It was a gigantic structure, 804 feet long, 264 feet wide, and 193 feet high (inside dimensions). It was the largest open-floor, that is to say without internal supports, building in the world.

The ZR-1 would be assembled inside. Its girder work would be fabricated at the Naval Aircraft Factory in Philadelphia and brought by truck or rail to Lakehurst. The pieces would consist mainly of triangular-shaped duralumin girders fourteen inches high. They had been the trademarks of recently defeated Germany's zeppelin constructors.

Before Philadelphia could begin stamping out the pieces and riveting them together, a source had to be found for the duralumin. It was a very strong and light alloy of aluminum, copper, and other elements. Germany had produced it but the United States had little capacity to do so. After some arm-twisting by the

ZR-1 under construction at Lakehurst. *U.S. Navy Photo*

Navy, the Aluminum Company of America, recognizing the potential of the metal for other aircraft as well, agreed to furnish it.

The ZR-1, which the Navy was anxious to build itself and not purchase overseas—it wanted to learn rigid airship design and construction by doing it—was an American-built copy of a German naval zeppelin from the war years. The L-49 had been shot down in France virtually intact. The French had made technical drawings of it and supplied them to the Allies. Except for certain changes, such as adding a mooring assembly to the bow, the ZR-1 followed the L-49's plans.

Transverse duralumin frames and rings, connected one to the other by longitudinals, typified the construction of a zeppelin hull and were embodied in the ZR-1. They gave the ship its shape, strength, and rigidity. Inside, through the length of the hull, twenty gas cells were arranged. These were cotton bags lined with intestinal membranes from hundreds of thousands of oxen. In today's world of synthetic fabrics and plastic films, it seems to have been inane to use a quarter of a million animal intestines to make an airship gastight. But it was the 1920s. These membranes, called "goldbeater's skins," were the best material to use. They were called thus because they were used by goldsmiths to protect gold when they were beating it into leaf.

The cells had maneuvering valves located at their top where gas pressure was highest. They were actuated, opened, and closed by lines pulled from the control car. Pressure-relief valves were at the cell bottoms. They opened automatically when the gas pressure inside—helium was the gas that would be used—became too high.

The outer cover of cotton cloth was laced to the frame, and painted with aluminized aircraft dope to tauten it and make it waterproof. The dope gave the cigar-shaped craft a bright silvery appearance. The color was intended to reflect the sun's heat to reduce its effect on the temperature, volume, and lift of the gas inside.

Propulsion was by six three-hundred-horsepower Packard engines, five in outboard engine cars ("eggs") and one in the after end of the control car. The ship was designed to do sixty-eight miles per hour. However, being a replica of a "height climber" zeppelin of lightweight construction, it would be generally operated at lower speeds to avoid the fate of the R-38.

A helium repurification and storage plant was being built at Lakehurst. Working clerically at this task was an attractive young lady, twenty-five years old, with red hair and blue eyes. Her name was Joy Bright Little. She had been recently widowed.

After secretarial school, Joy Bright had enlisted in the Navy in World War I and been assigned to the Cape May air station. There she met Lt. Charles Gray Little. He had been in Newfoundland to oversee ground handling of the C-5 during its transatlantic attempt. His leg had been broken as he fell from the airship when it broke loose.

Charles Little and Joy Bright became engaged. He had orders to England and the R-38. She followed him overseas and they were married. Fourteen months later, the R-38 crashed. Lieutenant Little did not survive.

To help recover from her grief, Joy Bright Little found a job in a newly established office within the Department of the Navy. It was the Bureau of Aeronautics (BuAer) created in September 1921. The president had appointed Rear Adm. William Adger Moffett its first chief. Moffett would prove an exceptional chief of BuAer. A soft-spoken southern gentleman, he was liked and respected by members of Congress who found him quite different from the Army's brash Col. (formerly brigadier general) "Billy" Mitchell. Moffett was the right man to thwart Mitchell's efforts to take air power away from the Navy and make it a part of an independent air force. In this and in his view of the future, he was, indeed, "the father of naval aviation."

Joy Bright Little liked working for Moffett in Washington but wanted to be closer to Lighter-than-Air, the Navy's balloon and airship service. She had lost her husband in airships but not her enthusiasm or devotion to their cause. Hearing that jobs were available at Lakehurst, she applied and went to work for Lt. Zeno Wick building the helium plant. It was at Lakehurst that she would meet her second husband. He was Lt. Cdr. Lewis Hancock Jr., a former submariner who was in training for duty on the ZR-1.

On 20 August 1923, the airship was floated in the Lakehurst hangar. On 4 September, it was undocked and moved by 350 ground crewmen onto the field. Its silvery hull, 680 feet long, 78 feet wide, and 2,235,000 cubic feet in volume, made a magnificent sight as it sat there, its six giant wooden propellers slowly turning over.

The dirigible was "weighed off" and ballasted to make sure it was light enough to lift off. Its captain, leaning out of a window in the control car that hung from the forward underside, megaphoned, "Up ship!" The ground crew let go, helping it with an upward shove. The aerial giant was free. The Packard engines roared to life. "Up elevator" was ordered. That would put the tail down and the nose up. The ZR-1 began moving slowly forward and upward.

Other flights soon followed, including hand-waving trips over the eastern states to show off the ship to the millions on the ground wanting to see it.

On 10 October, the ZR-1 was commissioned and christened USS *Shenandoah,* an Indian word meaning "daughter of the stars."

During its initial flights, the *Shenandoah* had three "captains."

McCrary, also Lakehurst's commanding officer, was the ZR-1 commander. He had become the Navy's first balloonist in 1915 and pilot of the airship DN-1 in 1917.

Cdr. Ralph Weyerbacher of the Navy's Construction Corps had supervised the ship's construction. He was on board in charge of seeing that it passed its builder's trials.

The third was a German civilian, a former Zeppelin Company pilot, who had been hired by the Navy to instruct it in how to fly its new airship. Anton Heinen, short, balding, with a red goatee, and feisty, was the only person on the ZR-1 who had any experience commanding a rigid airship.

One night, Heinen saved the *Shenandoah*.

To avoid cross-hangar winds while undocking and docking, a 160-foot-tall mast had been erected at Lakehurst to enable the airship to "moor out."

The evening of 16 January 1924 was one of gale winds that gusted to seventy miles per hour. Their violence ripped the fabric from the upper fin and wrenched

ZR-1, named the *Shenandoah,* moors to the Lakehurst high mast. *U.S. Navy Photo*

the *Shenandoah* from the mast. With its forward-most gas cells deflating, it broke away, leaving some of its mooring assembly hanging on the masthead.

Its upper fin was in tatters. There was a gaping hole in its bow. Punctured gas cells were leaking helium. The airship began drifting tail-first with the gale, barely clearing the tops of the New Jersey pines.

Twenty-four men, half the usual crew, were on board. Heinen was one. He quickly stepped forward and took over.

As the *Shenandoah* drifted north toward New York City, he was able, by use of engines, ballast (rigids used water, not sand), and careful steering, to bring the ship, limping, back to the air station the next morning.

While he was struggling to control the floundering airship, Heinen was faced in the control car by a junior officer who thought he knew better what to do. After putting up with him briefly, Heinen ordered the inexperienced upstart out of the car. He did so, according to the stories told, with a kick to the officer's behind.

The *Shenandoah*, overhauled and repaired, was flying again in May. While the ship was out of service, its sixth engine was removed from the control car and replaced by a radio shack. Its command structure was changed. McCrary was transferred to non-lighter-than-air duty. Weyerbacher, now that the ship had completed its trials, was no longer part of its operational picture. Anton Heinen's services were no longer required. In their stead, there was a new commanding officer.

Lt. Cdr. Zachary Lansdowne, Naval Academy Class of 1909, had served in destroyers before he applied for flight training. At the time, married officers had to obtain their wives' approval before they could be accepted. Lansdowne's said no. She thought flying too dangerous. Shortly afterward she died, leaving him a son. He reapplied in December 1916 and was ordered to Pensacola for instruction in heavier-than-air craft. A year later, he was sent to Akron for instruction in airships.

Assignment followed to England for more training in British airships. There, he so impressed his hosts that they invited him to join them on the transatlantic flight being planned for the R-34. He was on board that rigid airship as a U.S. Navy observer from 2 July to 5 July 1919 from East Fortune, Scotland, to Long Island, New York. This made him the first American to fly the Atlantic nonstop, for which he was awarded the Navy Cross.

Lansdowne's 108-hour flight to America was followed by duty at the Navy Department in Washington and that by a tour as commanding officer of the naval air station at Akron. When it was closed in 1921, he was returned to Washington to assist Admiral Moffett in BuAer.

In Washington, he met Margaret Selden Ross, nineteen years old and a post-debutante. "Betsy" came from a Navy family and was socially well connected. A popular brunette, she was an ideal "date" for the handsome six-foot-tall widower. "Zach" Lansdowne and "Betsy" Ross were married in Washington's National Cathedral at year's end in 1921.

From Washington, Lansdowne was ordered to Germany, to the American Embassy in Berlin. A new rigid airship, a German zeppelin, was under construction for the Navy at Friedrichshafen. Two of Lansdowne's shipmates, Lt. Cdr. Garland "Froggy" Fulton and Lt. Cdr. Ralph "Horse" Penoyer, were also in the country to establish an inspector of naval aircraft's office there.

At the embassy, Lansdowne handled whatever aeronautical matters came up. The Navy's participation in the Gordon Bennett international balloon races was among them. Lansdowne assisted Navy balloonists in their preparations for the 1922 and 1923 events in Geneva and Brussels. Lansdowne was present at these races including the one in 1923, which saw five men killed and a half-dozen balloons destroyed or wrecked.

After two years in the German capital—for part of the time the naval attaché was William F. Halsey—the Lansdownes returned to Washington. Admiral Moffett wanted him back to plan for the use of the *Shenandoah,* already flying, and the ZR-3, the German ship soon to be delivered. Moffett had in mind sending the *Shenandoah* to the North Pole!

A few months later, Admiral Moffett ordered Lansdowne to Lakehurst to relieve McCrary and be the *Shenandoah*'s commander. "Zach" was enthusiastic about his new assignment. He wanted to downplay the ship's public relations role and demonstrate its potential as a long-range naval scout.

In August 1924, he had his opportunity to work his command with the surface Navy. An oiler, the USS *Patoka,* had been converted into a tender for rigid airships. The *Shenandoah* was to moor to it in Narragansett Bay.

On its stern, the *Patoka* had a mooring mast like the one at Lakehurst, although seventy feet shorter. Off Newport, Rhode Island, on 8 August, the airship approached the *Patoka* as it steamed slowly into the wind. When the dirigible's mooring cable went slack, it jumped a sheave. The problem was righted and the airship winched down and to the mast top. Thus was demonstrated for the first time how a rigid airship could operate from a surface ship.

Back at Lakehurst, an infusion of Texas helium enabled *Shenandoah* to show again its naval usefulness. Flying through heavy rain in August, it searched for and found the battleship *New York*.

Then came *Shenandoah*'s longest flight: Lakehurst to Fort Worth, Texas; Fort Worth to San Diego, California; San Diego to Tacoma, Washington; and return

Shenandoah moored to the *Patoka*. U.S. Navy Photo

via the same route. Masts were erected for it at these stopover points. The round trip began on 7 October and ended on the twenty-fifth. The *Shenandoah* arrived home at Lakehurst to find a strange new airship waiting for it in its hangar.

In the division of spoils following the war with Germany, the United States was to have received two of the surviving naval zeppelins. It did not. A number of these ships, including those intended for America, were destroyed by their crews to keep them out of Allied hands.

As compensation, the United States pressed for a replacement zeppelin to be built and for the German postwar government to pay the bill. Britain and France, not wanting America to have a newer, bigger, and better zeppelin than they had, strongly opposed the idea. There was much bickering, some of it bitter. The United States reminded Britain that it had sold it a faulty airship, the R-38, which had killed the core of the U.S. Navy's rigid airship personnel.

Feelings became so aroused that even the American secretary of state, Charles Evans Hughes, felt it necessary to intercede. Britain and France finally relented but added an ironic twist. The United States could have its zeppelin but

Admiral Moffett, BuAer chief, and Commander Lansdowne (left to right) ride a lift (a balloon basket) to the top of the mast where the *Shenandoah* was moored at North Island, San Diego. *U.S. Navy Photo*

could operate it only for civil purposes. An insult to impose upon a military service of a former wartime ally!

The United States accepted the condition.

Out of this came a 658-foot-long, 90-foot-wide, 2,760,000-cubic-foot zeppelin, bearing the builder number LZ-126 and the U.S. Navy's designator ZR-3 (Lighter-than-Air, Rigid, #3).

Luftschiffbau-Zeppelin made it the best airship it was capable of, expecting it would be its last. Upon completion of the ship, the company's hangars were to be torn down by order of the Allied powers.

The ZR-3 and the ZR-1 were alike in many ways, both embodying typical zeppelin construction features, but there were a few differences: The ZR-3 had fourteen goldbeater's-skin-lined gas cells; the ZR-1 had twenty. The ZR-3 was powered by five four-hundred-horsepower German-built Maybach engines, whereas the ZR-1 had six (later five) three-hundred-horsepower Packards. The ZR-3 was fatter, providing better resistance to bending, and its control car was enclosed and built flush against its underside. That of the ZR-1 hung from the hull by wires and struts.

The ZR-3 was flown from Friedrichshafen to Lakehurst, arriving on 15 October 1924. Quickly it was emptied of its flammable hydrogen, which was released into the air. Then it was hung, empty, from the hangar overhead and shored up from the floor to await the *Shenandoah*'s return. There was not yet enough helium to inflate two large airships at the same time. The older and the newer ship would have to take turns sharing their gas.

While waiting for its helium, the German ship, which arrived with no markings, was painted with Navy colors. In addition to a red-dotted white star in a circular field of blue under its nose, on its back, and on either side of its stern quarter, "U.S. Navy" was spelled out in large black letters amidships on either side. "Los Angeles" was added in smaller letters of blue directly forward of and below the horizontal fins. Rudder and elevator surfaces were given red, white, and blue stripes.

The helium transfusion from the *Shenandoah* was made. On 25 November, the ZR-3, commanded by Capt. George W. Steele, flew to the Naval Air Station, Anacostia, Washington, D.C., there to be christened by the nation's first lady, Mrs. Calvin Coolidge. She named it *Los Angeles* and broke a bottle of water, purportedly from the river Jordan, against the control car rail.

While on the field at Anacostia, the airship grew restless and increasingly difficult to control. The sun was warming the helium and generating more lift. The ceremony had to be speeded up. So did the president. A surprised Coolidge heard an authoritative Navy voice barking at him: "Step lively! Step lively! Let's keep moving!" He must have been even more startled to hear another blurt out: "Make way for President Harding!"

Newly named, the USS *Los Angeles* returned to Lakehurst to begin a series of flights that included operating with the *Patoka*. It would continue to fly during the 1924–25 winter and into the spring, when it would be time to change helium again.

Refilled with helium, the *Shenandoah* began flying again. In September, it was to make a weeklong, hand-waving, public-relations flight to the Midwest, visiting St. Louis, Des Moines, Minneapolis, and Detroit. The chief of naval operations had wanted the flight to be made early in the summer but Lansdowne had pointed out the disadvantages. The ship would have less lift in the warm and less dense air. There was also the hazard of thunderstorms. If he had to make the flight, he preferred the second week in September. The Navy insisted. It had to be the first week to satisfy certain state fairs along the route. On 2 September 1925, the ship departed Lakehurst to carry out these orders.

USS *Los Angeles,* maximum speed seventy-four miles per hour, early in its naval service. *U.S. Navy Photo*

It never returned.

In the early morning of the third, about 4:45 AM, the *Shenandoah* was entrapped by a developing storm near Ava, Ohio. Its long, thin, pencil-like hull was thrown up, down, and around by vertical currents that broke it in two. The after half settled to the ground a third of a mile away from the point of break-up. The forward section was free-ballooned while those in it dropped water ballast and valved gas. It landed twelve miles away. When the supports holding the control car failed, it broke loose and fell to the ground, killing those in it. They included Lansdowne and his executive officer, Lt. Cdr. Lewis Hancock Jr. In all, fourteen died who had been in the control car, in the hull, and in two engine cars that broke away. There had been forty-three on the ship.

It was to have been Zachary Lansdowne's final flight as the *Shenandoah*'s commander. He was due for sea duty to qualify for promotion. He left Betsy, their two-year old daughter, Peggy, and the family dog, Barney, who inexplicably cried all the night his master was killed.

Joy Bright Little Hancock had lost two husbands in Navy dirigible disasters. She remained Navy through and through. A WAVE (Women Accepted for Volunteer Emergency Service) officer during World War II, she later became assistant chief of naval personnel for women with the rank of captain. Navy destroyers would be named for both her husbands.

SIX

Thomas Greenhow Williams Settle

On the morning of the fourth of July 1924, Lt. T. G. W. "Tex" Settle arrived at Lakehurst for duty. He had just graduated from Harvard University's Cruft High Tension Laboratory as a specialist in aviation communications engineering. His orders from the Bureau of Navigation, later the Bureau of Naval Personnel, were for him to become the air station's communications officer and communications officer for the *Shenandoah* as well.

"Tex," so named by his Naval Academy classmates—he had graduated in 1918, second in his class—had entered the Navy from Galveston where his father, a career Army officer, was stationed.

When he reported to Lansdowne the next day, the *Shenandoah*'s new skipper made it clear to his new communications officer that he wanted the best possible communications between the airship and Navy surface units. To provide this, Settle would have a brand-new radio shack that had replaced an engine in the control car. The dirigible transmitted in the 250–600 kilocycle range and at 3,332 kilocycles. It received in the high, low, and intermediate frequency bands. For an antenna, it trailed a 450-foot wire that was weighted at its end by a fifteen-pound lead "fish."

Settle was enthusiastic about it all, including Lakehurst's cavernous hangar, its skyscraping mooring mast, and its great open landing field. He was fascinated even more by the prospects that the *Shenandoah,* with him on board, would be casting off for the North Pole. (The ship's polar flight would never take place. Its chances of success were too marginal for President Coolidge, who ordered it cancelled. Not until 1926 would anyone reach the North Pole by air. Richard E. Byrd claimed to have done so that year by plane from Spitsbergen. His claim has since been questioned. So has Robert E. Peary's that he was first to discover the Pole [by sledge] in 1909. On 12 May 1926, three days after Byrd's flight, the airship *Norge*—with Norwegian Roald Amundsen, American Lincoln Ellsworth, and Italian Umberto Nobile on board—overflew the North Pole en route from Spitsbergen to Alaska. Amundsen, who discovered the South Pole, may also have discovered the North!)

While "Tex" enthused, his bride of several weeks mused about how she would keep house for him. Fay Brackett, with hazel eyes and chestnut hair, had met him at the Cruft Laboratory where he was under instruction and she was working. They were married in early June in her hometown of Arlington, Massachusetts. Fay was accustomed to New England. She found that Lakehurst took some getting used to.

There was a town nearby by that name but it had only a thousand people. Other than being the site of the new dirigible base, its claim to fame was that it was a stop on the New York to Atlantic City run of the Central Railroad of New Jersey.

The U.S. Naval Air Station, Lakehurst, New Jersey, while not a "hardship station," came close to being one. It was remote. Visitors often likened it to an island in a sea of pine trees. (Locals who inhabited the woods were known as "pineys.") The soil was barren. Navy wives, who tried to grow flowers around their quarters, found that only marigolds flourished.

The families of junior officers, such as Settle, were allotted a two-story duplex apartment with some government-provided furnishings: a divan, rocking chair, dining room table with straight chairs, and buffet. They were dark wood and in the then-popular mission style. A large living and dining room, pantry, and kitchen were on the first floor, a bath, good-sized bedroom, and a smaller one on the second. Each apartment had a gong that banged out a summons whenever the *Shenandoah* was coming or going.

Much of the air station was unpaved. The same was true off base. Lakehurst people were used to driving on dusty washboard roads.

Navy families thanked heaven for the station's commissary. For other than food, they shopped mainly in Philadelphia and New York, each sixty miles away. They preferred Philadelphia, with its John Wanamaker department store, even though to get to it, you had to drive to Camden and ferry across the Delaware River. If you wanted to go to New York, you drove or could take the Jersey Central. But the railroad went only to Jersey City where you got off to take a boat across the Hudson River.

Isolated military bases, of course, make their own social life. Lakehurst was no exception. But there the social life was somewhat divided. The air station

The Naval Air Station, Lakehurst, in the late 1920s and early 1930s. *U.S. Navy Photo*

families had their own, and the airship families theirs. Everyone was friendly, but the two groups went their own ways. Perhaps this was due to the fact that Lakehurst had no officers club. Entertaining was done at home or in a large room in the bachelor officers' quarters.

Prohibition being the law of the land, the air station was "dry." Local bootleggers, working the nearby Atlantic shore, made sure that anyone who wanted liquor would have it. Lakehurst's favorite drink was "tiger's milk," a gin-based concoction.

Settle, after squaring away his "shack" and becoming familiar with the radio capabilities of the *Shenandoah,* accompanied it on its flights, except for its trip to the West Coast. A fleet communications specialist was not needed on that overland route. He would miss being on the airship when it was lost by meeting radio personality Graham McNamee in Chicago to arrange for a broadcast in flight.

After the ZR-3, the future *Los Angeles,* arrived, Settle was given the additional duty of being its communications officer as well. This was feasible because helium production had not yet been able to supply enough of the gas to inflate both ships at the same time. On whichever one was flying, "Tex" Settle would be the communicator.

One of his flights on the *Los Angeles* was interesting because its purpose was to observe a solar eclipse off New England. It was conducted on 24–25 January 1925 in air so cold the crew wore knitted pink "long johns," there being no heat in the ship.

It was that winter that Settle observed an incident that could have seriously damaged both the *Shenandoah* and *Los Angeles.*

On a cold morning, with ice on the ground, the hangar doors were opened to take the "L.A." out onto the field. It and the *Shenandoah* had been hangared side by side, the latter suspended from the overhead to await its next transfusion of helium.

As the *Los Angeles* began to emerge tail-first, a cross-hangar wind took hold of its stern and began to blow it across the doorway and into the other ship. The ground handlers could not halt the drift. Slipping and sliding on the ice, they could not control the dirigible. It seemed determined to crash its stern into the *Shenandoah*'s. Officers in the control car shouted encouragement to the handlers, apparently unable to think of anything else to do.

There was something else to do, and Hans Flemming, a captain from Luftschiffbau-Zeppelin, was the one who did it. Flemming had been on the ZR-3 during its delivery flight from Germany and had remained in the States temporarily to instruct the Navy in its operation.

From the way things were going as the *Los Angeles* was steadily coming closer to the *Shenandoah,* the situation was out of control. The Navy, in its inexperience, was unable to cope.

Without saying a word, Flemming stepped up to the engine telegraphs, which were located forward in the car over the starboard windows. "Ahead full," he rang up. The Maybach engines responded. The *Los Angeles* flew itself out of the crosswind and back into the hangar. Settle would never forget what he had seen: a superb example of the value of experience!

In mid-1925, the *Los Angeles* was deflated of its gas and laid up for overhaul. It was the *Shenandoah*'s turn to operate. When it was destroyed in September, the "L.A." would take its place.

Settle had a sense of initiative and personal drive that had to be seen to be believed. He was a package of restless energy, then and later on through life. (At age eighty, he was still bounding up and down stairs two at a time.) To anyone with his lively temperament, sitting cooped up in a radio shack while his brother officers were piloting balloons and airships was not to his liking. He arranged to be given lighter-than-air flight training on the *Los Angeles.* It was on 19 January

1927 when, "having fulfilled the conditions of the United States Navy Department," he was designated Naval Aviator (Airship) Number 3350.

"Tex" loved to fly and particularly he liked the challenge of free ballooning. Shortly after earning his wings, he checked out one of Lakehurst's small nineteen-thousand-cubic-foot balloons. He intended to fly it as far as he could. The wind was toward the northeast. When he landed twenty-one hours and thirty minutes later, he was at Lisbon Falls, Maine, having traveled 478 miles. Had he carried a barograph, which recorded altitude versus time, the FAI might have registered his flight as a world record for that size balloon.

Navy balloon racing had, meantime, fallen to a low. There had been national balloon races at San Antonio, Texas, and St. Joseph, Missouri, but Navy pilots did not participate.

The Gordon Bennett of 1924 was held 15 June at Brussels. DeMuyter was again the winner. He was opposed by fifteen competitors. He landed in Scotland after crossing the English Channel. His victory, the third in a row, automatically retired the trophy. The aero club of Belgium provided a replacement, enabling the Gordon Bennett races to continue. It was very Belgian, made of metal mined in that country and paid for by popular subscription.

No one could have imagined, as the entries rose into the sky, from Brussels again, in 1925, that one of them would land on the deck of a steamship in the Atlantic.

At night!

It was Ward Van Orman's *Goodyear III*. He and Carl K. Wollam had been the winners of the American "nationals" earlier in the year. Van Orman had not only won, he also received a trophy given by Paul W. Litchfield, a Goodyear Tire and Rubber Company executive. It was the first time it had been awarded. The Litchfield Trophy would continue to be given as long as the "nationals" were held.

Van Orman and Wollam were airborne at Brussels about six o'clock that evening, at first drifting toward Spain. When the wind veered toward the west and north, the coast of France came into view. For some time, they moved along it, "Van" expecting he would be able to make a landing near Brest. But the winds would not cooperate. He went to twenty-five thousand feet in search of one that would bring them back to land. He found none. Increasingly they were being taken out over the Atlantic.

A difficult situation became a critical, then a desperate, one. There was a receive-only radio on board, one of the first ever in a balloon race. Its bulletins offered no hope for a change in wind. Van Orman told Wollam not to lose heart. "We have a good chance of being picked up by a ship," he said.

Wollam didn't think so. He began searching for the bottle of cognac that he knew was in their provisions. He found it and began working on it. Its effects made themselves felt.

Convinced he was going to die by drowning, Wollam decided to end his life in a different way. There was a Very pistol in the basket for shooting signal flares. He took it, aimed overhead at the bag full of hydrogen, and fired.

He missed!

A mightily worried Van Orman tried to calm his companion. Thereupon, "Wollie" said he was going to throw himself overboard to rid the balloon of his weight and give "Van" a better chance to survive.

Before Wollam could do this, a bright light was seen on the inky-black surface below. It was from a small German freighter, the *Vaterland,* on its way from Cairo to Rotterdam. *Goodyear III* was two thousand feet up. With a flashlight, Van Orman spelled out in Morse code that he intended to land on the ship. Its German captain, Rudolf Nordman, fortunately knew some English. He read the message and understood. He turned on all the ship's lights that he could. He told his helmsman to steer for the balloon.

Balloon and vessel approached each other, one drifting in the air at forty-five miles per hour, the other steaming at seven. To slow himself down, Van Orman threw out a sea anchor. This was a canvas bag with a hole in the bottom. Dragged through the water, it reduced his drift to a manageable speed.

Airmanship and seamanship brought the two craft together. The balloon ended up against a rail on the *Vaterland*'s forward deck. It was quickly pulled completely on board by the German crew. Van Orman ripped the bag to empty it of its hydrogen.

Never before in the history of ballooning had there been a landing like this! It was a date to remember: 9 June 1925.

Van Orman and Wollam had flown farther than any other balloonists in the race. They didn't win. They were disqualified.

FAI rules required that a winning balloon descend on land. If one came down at sea, it would be disqualified. It was ruled that *Goodyear III*'s landing on a ship, even though the landing had been a controlled one, was a landing at sea. Thus, Van Orman and Wollam were deprived of victory. The judges then turned around and awarded it to a Belgian named Veenstra who had not only splashed down into the sea but also had to be rescued from it some hours later.

The American team protested but without success. When they left for home, they did so hurt, disappointed, and disillusioned. Fifty years later, Van Orman would write: "The memory of that unfair decision never has grown dim." He was determined to return and win the damn race.

In 1926, when he arrived for the national race at Little Rock, Arkansas, he had a new aide, Walter Morton, a stolid and dependable type. They won without difficulty, bettering the runner-up by more than two hundred miles. In the Gordon Bennett that followed from Antwerp on 30 May, they made it to Solvesborg, Sweden. A telegram informed them they were the winners. They had flown twice the distance of their closest competitor, the great Ernest DeMuyter.

Having won the 1926 Gordon Bennett, Ward Van Orman and Walter Morton were entitled to be in the next to defend their championship. They didn't have to enter the national race, but they wanted to. It was held in Akron, Ohio, on 30 May 1927. Through heavy rain and thunderstorms, they covered 718 miles to land in fog at Hancock, Maine. When they debarked from the basket and looked around, they found they had set down just 150 feet from the Atlantic Ocean. They had been first, civil pilot Edward J. "Eddie" Hill second, the U.S. Army's Capt. William E. Kepner third.

That year the Gordon Bennett originated on 10 September in the Detroit area at a flying field made available by Henry Ford. Hill was the winner, an entrant from Denmark second, Van Orman third. When the Goodyear balloon came down at Adrian, Georgia, a crowd gathered that tried to pack up the bag. They undid the wrong ropes, letting it slip out from under its net. When last seen, it was disappearing into a thunderstorm that had forced the crew to land.

There had been quite a few balloon races, national and international. What was happening, meanwhile, at Lakehurst?

The *Shenandoah* had been wrecked. The *Los Angeles* had resumed flying.

Navy speed flyer, Al Williams, had visited. "Passed through" was a better way to put it, inasmuch as he used the occasion to fly his plane through the hangar. (He had first checked to be sure there were no airships or cables hanging down inside.) Then, as a kind of grand finale, he taxied over to where automobiles were parked and proceeded to put his propeller through the roof of one of them.

Also there was the man who showed up wanting to jump over the hangar. In the 1920s, balloon jumping was considered great sport. Wearing a harness with a small balloon or two attached to its back, daredevils would leap into the air, enjoy the feeling of buoyancy, and let the wind take them where it would. No one remembers now whether the man was successful in his jump, or whether he was even allowed to try.

In 1927, the *Los Angeles* performed a feat unequalled by any rigid airship, before or after. It stood on its nose.

It was 25 August, a hot mid-summer's day at Lakehurst. The "L.A." was riding peacefully to the "high mast," the nose cone dangling from its bow locked into a cup at the top of the tower.

The officer who had the command watch was "Tex" Settle. Always one to "keep a weather eye," he had been observing small white clouds moving toward the air station from the shore seventeen miles to the east. These were a sign that a sea breeze was in the making. Increasingly concerned, he telephoned down to the captain, Lt. Cdr. Charles E. Rosendahl, at the base of the mast, urging that the airship cast off. Rosendahl saw no need for it.

The ocean's cool and denser air reached the dirigible's stern and engulfed it, increasing its lift. The force of the breeze, striking its underside, shoved the tail upward. The *Los Angeles* began to pitch up by the stern, to "kite" as airshipmen would say, while taking on an ever-increasing angle.

Settle could see and feel the ship's movement. He ordered water pumped up the mast and then to the stern of the ship to make the tail heavier. He sent crewmen on board rushing aft to help. They were known as the "galloping kilos," being ordered back and forth along the keel to adjust the distribution of weight and, therefore, the airship's trim.

Despite these measures, the *Los Angeles*'s stern was still rising. It reached 20 degrees . . . 55 . . . 70. . . .

Loose gear, tools, and the like were falling through the airship's length and collecting in its bow. Gasoline and oil were spilling from tanks and running down and staining the outer cover. A terrific crash was heard from the galley as the chinaware fell to the deck.

Settle looked out the front windows of the control car. He was looking straight down! He was walking on those windows!

Eighty-eight degrees.

About this point, the tail stopped rising. Slowly the "L.A." rotated about its nose, changed heading by 180 degrees, and started coming back down.

As it did, Settle had a different and contrary problem on his hands. How to stop the stern, loaded with water ballast as it was, from descending too far and striking the ground. He dumped ballast aft and ordered the "kilos" forward. The ship slowly leveled off and resumed its normal riding-to-the-mast attitude. The lower fin did not strike the ground. Miraculously no major damage had been done to the airship. Luftschiffbau-Zeppelin had built it well.

It was now obvious the high mast had to go. Rosendahl had a "stub mast" built that was sixty feet high. When moored to it, the *Los Angeles* lay practically on the ground. To hold the stern down and allow the ship to vane with the wind, its rear engine gondola was secured to a riding-out car that rolled as the ship changed heading. Later this stern handling car would travel on a circular railway track. The stub mast made it possible to board and to on-load and off-

On 25 August 1927, the *Los Angeles* stood on its nose! *U.S. Navy Photo*

load the ship through the control car and hatches along the bottom of its hull. Otherwise, on the high mast, 160 feet in the air, the dirigible was accessible only via the top of the mast.

    Lt. Cdr. Charles E. "Rosie" Rosendahl (Naval Academy Class of 1914; commanding officer, USS *Los Angeles,* 1926–29) was tall, blue-eyed, handsome, articulate, and fast becoming the Navy's and America's foremost champion of airships. Before commanding the "L.A.," he had served on board the *Shenandoah* as navigator and had been in the ship when it was destroyed in September 1925.

Rosendahl would have lost his life that night, along with the others in the control car, had not Lansdowne sent him up into the hull to prepare to lighten the ship by dropping fuel tanks. Moments after he left the car, it was torn loose.

The demands "Rosie" made on his officers and his attitude toward their wives—he didn't believe women belonged around airships—won him respect but no popularity contests. But his devotion to lighter-than-air flight and his professionalism resulted in one of the most noteworthy careers in naval aviation history. He would be one of the first Navy airmen enshrined in the Hall of Honor at the National Museum of Naval Aviation in Pensacola.

The record of the ZR-3, the *Los Angeles,* was no less notable. While in service, it would make 331 flights—more than 100 of them under Rosendahl's command—before it was decommissioned and retired in 1932. It never had a major accident. It never killed anybody. One of its officers, George F. Watson, used to say: "A good ship makes its own good luck."

The *Los Angeles* was and did.

SEVEN

ANOTHER NAVY RACE

After a three-year hiatus, the Navy took up balloon racing again on 30 May 1927 in the "nationals" at Akron, Ohio.

There were fourteen entries, including Navy balloon 3-97 crewed by Lt. T. G. W. Settle and Chief Boatswain's Mate George N. Steelman. Their thirty-five-thousand-cubic-foot sphere left the ground at 5:52 PM from the Cleveland Speedway. (Navy balloons were in three sizes: nineteen thousand, thirty-five thousand, and eighty thousand cubic feet. The latter were standard volumes for national and international balloon races.) They had to evade nearby thunderstorms.

Next morning Settle and Steelman were over upper New York State, drifting toward the St. Lawrence Valley and Nova Scotia with a wind of about twenty miles per hour.

At 9:45 AM, while under a layer of stratus clouds, they heard a deafening clap of thunder from directly above. A heavy, cold, rain began pouring down as the air temperature dropped from 60° to 35° Fahrenheit.

Settle valved hydrogen to get low to the ground. An ominous dark cloud passed overhead, its leading edge a rolling mass of turbulent air. It went by and the rain let up. The arrival of a second cloud brought another downpour.

Being almost out of ballast, they had to land. Settle brought 3-97 to earth at Pope Mills, New York, in a barnyard. A crowd gathered and enthusiastically helped roll the balloon up for shipment back to Lakehurst.

How the owner of a farm would react to an aerial invasion of his pea patch was one of the unknowns of ballooning. The bag, the basket, the net, the ropes, and the crowd of curious onlookers could make quite a mess of it. Pilots found it convenient to have some money on them to assuage the economic and other protests from those who felt their crops and fields had been violated.

The complaints came not from farmers alone. Owners of chickens, which had run wildly around, terrified by the sight of the strange bird overhead, were apt to claim that the Navy had disturbed the nesting and laying habits of their broods. Inasmuch as no one expected to be able to sue the Navy, whatever the balloonists could offer as recompense on the spot was usually acceptable. At Pope Mills, Settle and Steelman were welcomed.

They had made good 393 miles. Impressive, but not enough! Ward Van Orman and Walter Morton, in a new Goodyear balloon, had flown farther, landing and winning at Bar Harbor, Maine. Contestants Edward J. Hill with A. G. Schlosser placed second. The Army's Capt. William E. Kepner and Lt. William O. Eareckson were third. Hill and Schlosser would take first place and Van Orman and Morton third in the Gordon Bennett that followed at Detroit in August.

Settle had taken an Atwater Kent radio along to receive weather information during the race. The results were disappointing. The reports from commercial broadcasting stations, he said, were useless. Too much like newspaper weather forecasts. Not up-to-date enough. Not complete and detailed enough. He suggested that for future domestic races Navy Radio at Arlington, Virginia, issue weather information of the kind Navy pilots would need.

He wanted a Navy aerology (meteorology) team at future race launch sites. "Balloon racers can use all the weather help they can get," he pointed out.

He also thought the Navy should buy some new balloons: "The present ones are getting too old to be competitive. We fund racing planes, why not racing balloons?"

In 1927, "Tex" Settle had saved the *Los Angeles* when it performed its nose stand. In 1928, he was called upon to save another airship, the J-3, which he did by free ballooning it.

The Navy's line of nonrigid airships did not end with the H-1. There had been no I-class because "I" was too easily confused with "1." There had been a "J" however.

Two, the J-3 and J-4, were based at Lakehurst. Helium-filled and 193 feet long, they were serving as training ships, the Navy's lighter-than-air training having been transferred to Lakehurst from Pensacola and Hampton Roads.

Settle had dual duties at the station. He was engineering officer for the *Los Angeles* and commanding officer of the J-3. He often flew the smaller ship, a nonrigid, with trainees. They quickly learned to value his advice: "Fly the airship! Don't let the airship fly you!"

J-3 and J-4, Lakehurst's resident nonrigids, in the 1920s. *U.S. Navy Photo*

Four days after New Year's, 1928, he took off in the J-3 with two student officers, two enlisted trainees, and two air station machinist's mates. Their open gondola was suspended by cables from the bag to which they were attached by finger-shaped adhesive patches.

Just after passing Cassville, New Jersey, the J-3 suddenly lost both engines. Its 150-horsepower Unions stopped without warning. With no way on, the airship floated helplessly in the air, slowly rising and falling, and rising more than falling. Its hull took on an angle of 8 degrees down by the tail.

The wind, out of the west at fifteen miles per hour, began carrying the seven toward the coast, the ship gradually rising higher. The atmospheric haze dissipated as it did. The sun became brighter and more intense. Responding to its warmth, the helium increased in volume and in lift. By 3:30 PM, the ship was at forty-three hundred feet, drifting eastward at twenty-five miles per hour. It had crossed Barnegat Bay and was three miles offshore.

To prevent the bag from bursting as its pressure increased, Settle was forced to valve helium. He later described their predicament:

> It was necessary to valve helium. When the valve was pulled, the wire jammed, having jumped a sheave. It took some seconds before it could be cleared. Thereafter, all helium valving had to be done by reaching up and pulling the right wire. Helium was valved every few minutes, for a few seconds each time, to keep pace with the superheating and keep the rate of rise to a minimum. I did not want to valve more than necessary because we were close to maximum superheat [the sun would soon be setting]. When the superheat started on the downgrade, it would go fast and we had none too much disposable weight to use as ballast.

In other words: when the sun went down, the helium would cool and contract. Lift would be lost. The free-ballooning J-3 would descend and perhaps be forced into the sea. Valving to stay in the air was one thing. Getting the engines restarted was another.

The gasoline that fueled the Union engines came from a gravity tank that was supplied from storage tanks overhead. The gas line from the storage tanks was clogged with ice. To get the propellers turning again, gas had somehow to be supplied to that gravity tank.

Settle and crew figured out a way.

The J-3 carried a canvas bucket for use in picking up seawater for ballast. Opening a drain cock at the bottom of one of the storage tanks—with great difficulty, it turned out—fuel was drained, bucket by bucket, and hand-poured into the gravity tank. There was a lot of spillage and a lot of "goddams" but the seven managed to put enough into the tank to get the engines going.

Settle took the now-powered J-3 down to two or three hundred feet, where headwinds would be the weakest, and made for home.

That year, 1928, Rosendahl put lots of miles and hours on the *Los Angeles*.

On 27 January, for example, off Newport, Rhode Island, he landed it on the flight deck of the carrier *Saratoga*. The relative motion of airship and surface ship put a quick end to the experiment. The pounding that the rising and falling carrier deck gave to the floor of the dirigible's control car made Rosendahl decide to cast off. He did so and returned to Lakehurst, leaving his executive officer, Herbert V. Wiley, standing on the carrier's deck wondering how he would get home.

Never again would the Navy try landing a rigid airship on a carrier, using the mast-equipped tender *Patoka* instead. During and after World War II, Navy nonrigids would operate routinely from flattops.

The night of 3 March 1928 was a memorable one for the officers, crewmen, and ground handlers of the *Los Angeles*. The ship was arriving home to land after returning from a trip to Panama and Cuba. The air at Lakehurst was turbulent with winds as high as fifty miles per hour. The "tall mast" was still in use. As the *Los Angeles* approached to moor to it, one of the ship's yaw lines parted. Then its mooring cable snapped. Freed of these restraints, the ship rose to a thousand feet. Licking its wounds, it circled back to wait out a lull. During the night, it returned again and again to try to land.

After a while, about four o'clock in the morning, the wind abated. Rosendahl decided it was the time again to try to place the ship into the hands of the ground crew. It could no longer moor directly to the mast because its cable had given way. The "L.A." would have to be taken physically in hand on the ground and moved into the hangar.

The dirigible nosed down, its lines were caught by the ground handlers, and it sat quietly on the field. Suddenly, without warning in the early morning darkness, it was slammed broadside by a snow squall that began blowing it sideways. The cold air that accompanied the squall gave the ship greater lift. It struggled to rise and get away. As it was blown across the landing area, it was in the direction of trees along the station's perimeter. Three hundred ground crewmen strained to control it.

This was no time or place for an airship and Rosendahl knew it. "Up ship!" was his order. "Let go and stand clear!" In the commotion, not everyone could hear him. As the "L.A." took to the air again, its engines revving up as it did, an agitated officer rushed up to Rosendahl: "My God, Captain, we've carried men up on the rails!"

Six of the ground crew had been pulled into the air. Two were hanging on the after engine car rails, four on the control car rails. Those inside the control car reached out to grab the hands of those outside and help them get on board. One man was hanging by one hand, trying to support a fellow ground handler with the other. Seaman Second Class Donald L. Lipke had come on board during the brief time the ship was on the ground. He climbed out a window and made his way along the rail to help those clinging to it. He would be recommended for the Navy's Life Saving Medal. He was denied it, so the Department said, because it was intended to recognize only the saving of life at sea.

No one had been killed and no one was seriously injured. All the men on the control car rail were pulled on board. Of the two carried up by the aft engine gondola, one dropped off as he felt his feet leave the ground, dropped a short distance, and broke a toe. The other was able to pull himself back on board.

As it had when it stood on its nose, the *Los Angeles* demonstrated that it was, indeed, a lucky ship!

EIGHT

Lightning and Thunder

According to the *Pittsburgh Sun Telegraph,* 175,000 persons converged by car, truck, and bus on Bettis Field at McKeesport, Pennsylvania, on 30 May 1928. McKeesport, a suburb southeast of Pittsburgh, was hosting the National Balloon Race. The multitude created the largest traffic jam in Pennsylvania's history. Many who couldn't get into the people-choked field parked up to five miles away and from there watched the bags go up.

Settle was on hand, as was his aide from the year before, Chief Boatswain's Mate George N. Steelman. They were to crew one of the two Navy balloons in the race; Lt. J. H. Stevens and Lt. (jg) G. F. Watson were to crew the other.

Pilots, aides, and ground crews (enlisted men from Lakehurst) had been organized into a racing team with an officer-in-charge, Lt. Charles E. Bauch. Meteorologist Lt. Francis W. Reichelderfer was also a team member and on the scene.

Between races, "Tex" had been practicing his ballooning. He had made news when he landed on a rooftop in East Granville, Pennsylvania. He had made even more when he descended at North Hampton, New Hampshire. With him was Chief Radioman R. W. Copeland. They were testing a new radio receiver. Settle had been instructed to fly it as far as possible.

He did so to within three miles of the Atlantic Ocean. When he saw it coming up, he made a hasty landing in the town's public park. The balloon hit hard, bounced fifty feet into the air, and continued on its way along the tracks of the Boston and Maine Railroad until it became entangled in telegraph wires. The radio test was a success, Settle's and Copeland's reward a steak dinner from an impressed and generous local resident.

Starting lineup for the tragedy-laden 1928 National Balloon Race from Bettis Field. *Goodyear Tire and Rubber Co. Collection, University of Akron Archives*

The shirt-sleeved meteorologist is Francis W. Reichelderfer, himself a balloonist. Here the future chief of the Weather Bureau dispenses advice to Navy participants in the 1930 national race in Houston. *U.S. Navy Photo*

The people awaiting the takeoffs at Bettis grew anxious and restless. They knew the forecast was for thunderstorms. They wondered what it was like to fly a hydrogen-filled balloon in one.

Settle, in a thirty-five-thousand-cubic-foot bag, would do just that. Here, in his own words, is what he experienced that afternoon:

> At 1800 took off with 34 bags of sand ballast and approximately 163 pounds of instruments and equipment. Balloon was 30 lbs light with a full bag of gas.
>
> When several hundred feet above the field, we saw a bolt of lightning and heard heavy thunder to the north in a large and menacing cumulonimbus. The cloud had as its southern boundary a line running ENE and WSW. It was about four miles from us. Decision was made to go up immediately to try to get into a northwest airflow and, if possible, keep ahead or diverge from the storm. Two bags of ballast (70 lbs) were dropped during the next ten minutes to keep us rising. However the cloud was approaching us rapidly and, with the balloon at 4,500 feet, it overtook us and passed overhead.
>
> We started up rapidly, the variometer [rate-of-climb indicator] going hard over on the ascent side. At 8,300 feet, I valved two seconds. We leveled off at 8,400, and then started down at a violent rate.
>
> We were falling through air that was extremely turbulent. We had hail the size of peas, heavy rain, and snow flurries. The basket swung violently from side to side like a pendulum. It also spun in azimuth. We were hit by heavy gusts.
>
> There were lightning and thunder right on top of us. Rain and air were very cold. The lowest temperature recorded by our thermometer, which I kept in my pocket, was 34 degrees Fahrenheit.
>
> During our descent, we dropped fifteen bags of ballast. We came through the clouds at 3,000 feet and leveled off at about 2,000. Another balloon emerged through the cloud base to the northeast of us but we could not identify it and quickly lost sight of it in the rain and clouds.
>
> We found ourselves under a dense black cloud, the line boundary of which was several miles to the south of us. The line extended as far as I could see in either direction. There were areas of heavy rain along it to our right and left and heavy rain to the north. We had lightning and thunder on all sides. The rain in our vicinity temporarily eased, the wind then being out of the northwest about 30mph. For a few minutes, it was relatively smooth. But it was apparent that we were fast catching up with the line cloud.
>
> As we did, we started up again and began violently flying a "vertical circle," repeating the first one and reaching 8,200 feet. This second circle was more violent than the first. At times the cloud through which we were passing was so

thick we couldn't see the envelope above us. When nearing pressure height [the altitude at which the bag would be completely full] on the second rise, I valved about five seconds. Descending, we passed through the cloud base much as we had on the first circle and leveled off at 1,500 feet with nine bags of ballast remaining.

The cloud roof, under which we were flying, sloped downward toward its leading edge, which was well-defined. The air below and along this edge was in great turmoil looking like it was "boiling." Lightning and rain were heavy along it and astern of us. Under the cloud it was very dark, but it was light beyond it. It was like looking out into the daylight from a large, low, roof.

Once again we found ourselves overhauling the line cloud, this time in air that was relatively smooth. I searched the ground for a lee of some kind behind which we could lay to, but none was in our path.

When we reached the line cloud, we started up a third time but were able to check our ascent at 3,000 feet by valving.

When the vertical currents took us up for a fourth time, I was able to check our ascent just above the cloud base. Seeing a suitable landing place ahead, I valved to get us down.

We landed at 8:20 PM, a little more than two hours after takeoff, on the farm of Charles Stutzlager, one and a half miles south of Perryopolis, Pennsylvania. We used our last bag of ballast on landing and ripped the balloon. We were in the lee of high ground. It was raining heavily with much lightning and thunder.

All instruments and removable equipment were taken to Perryopolis and the balloon left in the custody of Mr. Stutzlager. During the night, vandals and souvenir hunters stole the valve, load ring, drag rope, rip cord, and all loose miscellaneous lines. They left the envelope, net, basket, and flotation pontoons intact.

(Pontoons, as flotation gear, were attached to one side and to the bottom of the basket. They were pioneered by Van Orman, one of the inventors of Goodyear's famed inflatable life raft.) Settle and Steelman were down and safe. Others in the race were not so fortunate.

The Goodyear balloon, with Van Orman and Morton, took a direct lightning hit. The bolt killed Morton instantly and stunned Van Orman. It set the bag on fire, except for a piece of it about twenty-five feet across, which parachuted up into the net, slowing the burning balloon as it fell three thousand feet to earth. That and the pontoons on the basket enabled Van Orman to survive impact with the ground.

Walter Morton and Ward Van Orman (left to right), in a photo taken the day of the 1928 national race. *The Lighter-Than-Air Society*

When he regained consciousness, he thought flies were buzzing in his face. Coming to, he realized they were raindrops. He was lying on the ground with his ankle broken. His friend and companion, Morton, was in the basket, dead.

Two other balloons fell victim to the lightning. An Army entry from Langley Field, with Lt. Paul Evert and Lt. Uzal G. Ent, was struck and set ablaze. Evert was killed in the basket. Ent stayed with him in the burning balloon, trying to revive him. He didn't stop trying and rode the basket all the way down. It landed in a river. Ent escaped with his life.

Carl K. Wollam and J. F. Cooper were flying the *City of Cleveland* when hit. Cooper was knocked out. Wollam stayed on board in an attempt to bring his partner to. Unable to get him into his chute and over the side, Wollam bailed out to lighten the balloon and improve Cooper's chances. He had waited so long, however, that he didn't have time to put on his own. So he wrapped it around himself and jumped from the basket. He landed safely. Cooper landed severely burned but did survive.

All but three of the balloons in the 1928 national race were overwhelmed by the weather and forced to the ground shortly after taking off. Settle and Steelman made good twenty-one miles. The other Lakehurst entry, with Stevens and Watson, flew seven.

It seemed impossible that any team could have mastered that weather, but three did. Two of them landed in Virginia, 261 and 248 miles away. One came down in Pennsylvania after 187 miles.

Capt. William E. Kepner and Lt. William O. Eareckson from the Army's Scott Field in Illinois were in first place. They made it to Weems, Virginia. They encountered the same violent weather. While fighting it, they jackknifed over a group of six high-tension lines. In addition, they collided with a railroad telegraph pole and snapped it off. It caught in the rigging. They used their feet and hands to push it loose so it would drop away. Attacked by rain and snow, they felt during the night like they were freezing to death. The balloon's ropes became so ice-encrusted that they crackled when they were touched or moved. Eventually they came to where Chesapeake Bay meets the Atlantic. There they landed. (Kepner and Eareckson would go on to win the Gordon Bennett from Detroit the following September with a flight of 460 miles.)

So ended the 1928 "nationals."

NINE

THE *GRAF ZEPPELIN*

A NEW RIGID AIRSHIP, A FOREIGN ONE, LANDED AT LAKEHURST ON 15 October 1928. Its name was *Graf Zeppelin*. It was completing its first transatlantic passenger- and cargo-carrying flight from Friedrichshafen, Germany.

The engineers who built the *Los Angeles* (to the Germans the LZ-126) had expected it to be their last zeppelin. The victorious Allied powers were intent upon destroying the Luftschiffbau-Zeppelin hangars. However, they had relented. The LZ-126 would not be the last zeppelin after all. Germany would be allowed to build airships again. The *Graf Zeppelin* was the result.

Dr. Hugo Eckener, an economist with a degree in psychology, had worked with Count Zeppelin over the years and become a zeppelin pilot second to none. Wearer of the count's mantle after his death, he had a goal of establishing a zeppelin service between Germany and America. Eckener raised the money by popular subscription, appealing to school children and adults alike. The Weimar Republic reluctantly provided the additional funds that were needed.

The size of the ship had been dictated by the clearances of the Zeppelin Company's largest hangar. The *Graf* would be 776 feet long, 100 feet wide, and 3,900,000 cubic feet in volume. It closely resembled its predecessor, the *Los Angeles,* except that its control car was larger and fitted with ten staterooms and a lounge for twenty passengers. The *Graf* had Maybach engines, totaling 2,650 horsepower, that it carried externally in five engine cars.

The Maybachs ran on gasoline, also on a gaseous fuel called Blaugas, the latter weighing only slightly more than air. Twelve of the ship's seventeen oxen-intestine-lined gas cells had an upper and lower half. The upper held hydrogen, the lower Blaugas.

Graf Zeppelin. U.S. Air Force

Consuming Blaugas had little effect on the static condition or buoyancy of the ship, whereas burning gasoline had considerable effect. The *Los Angeles* and the *Shenandoah* mounted condensers on their engine cars to recover the water in their exhaust gases and thus compensate for the weight of gasoline consumed. (Lightness brought about by gasoline consumption could, of course, be compensated for by valving, but in helium-filled airships, this was done only rarely and in emergencies because of the scarcity and cost of the gas.)

On its first crossing of the Atlantic in October 1928, with Dr. Hugo Eckener himself as captain, the *Graf Zeppelin* hovered on the brink of disaster. Encountering a severe cold front and unable to find a light spot in it, the *Graf* plunged through in black clouds that reached practically to the water.

As it entered the cloud wall, its nose was shoved downward. "Up elevator!" It recovered, taking on a bow-up angle of 15 degrees. It regained normal flight. Lightning flashed, rain fell in torrents, and water poured in everywhere. Then came the report that fabric on the port fin had carried away!

A party of five, which included Hugo Eckener's son, Knut, climbed out onto the swaying, wind- and rain-swept stabilizer to cut away the flapping fabric and tie down what remained. The airstream, too strong for them to work in, threatened to blow them off the ship. So the captain began a dangerous game. He would take the airship up to fifteen hundred feet, cut back the engines, and let it drift with no way on to give the men on the fin a chance to work. As it drifted, it also settled. When it was within three hundred feet of the ocean waves, Eckener would start up the engines and take the *Graf* back up to fifteen hundred feet. The process would then be repeated, the work party leaving the fin and withdrawing inside the hull while the dirigible was under power and climbing.

Lieutenant Commander Rosendahl from the *Los Angeles* was on board as a Navy observer. Eckener asked him to radio the Navy for help. Rosendahl wrote out the following: "Position lat 32N, long 42W, course for Cape Hatteras. Proceeding half speed about 35 knots airspeed on account of damage to cover of port horizontal. Effecting repairs as conditions permit. Request surface vessel proceed along our course and stand by. Request weather conditions to westward. In rain squalls at present."

This message would never have been received had it not been for an Italian steamer that picked it up and relayed it to shore. The airship was unable to contact the American mainland itself because, at its reduced speed, its wind-driven generator was unable to produce enough power.

From its very start, the flight had been front-page news the world over. As it proceeded, the headlines about it grew ever larger. The *New York Herald-Tribune* reported: "Zeppelin forced off course; wanders near Bermuda; ignores Navy calls." The *New York Times,* for perhaps the first time in its history, printed a large weather map on page one of its edition for Sunday, 14 October, with the words: "Zeppelin speeds on past Bermuda and may reach here before noon: damaged fin is repaired in the air."

The public, unaware that the *Times*'s prediction was premature, climbed into their cars and headed for Lakehurst "to see the Zep come in." The roads near the naval air station became parking lots. Residents of the town of Lakehurst rented out their driveways and even their yards to motorists looking for a place to stop and wait and watch. Hawkers sold hot dogs, chili, and popcorn. The area was littered with paper, wrappers, and boxes. The crowd that Sunday was estimated at two hundred thousand!

On the base, which was open to the public, the *Los Angeles* and the blimps J-3 and J-4 were in the hangar for viewing.

The station's communications office was keeping a vigil, waiting to hear from D-E-N-N-E, the airship's call sign. Its radiomen called every fifteen minutes to try to make contact.

The *Los Angeles* was "on the ready" to go to the *Graf*'s assistance. It was the only aircraft in America that had the range to reach it.

Meanwhile, the men on the fin were able to bring the damage there under control. They cleared away the flapping fabric that could have jammed the elevator. They rigged blankets where the fin joined the hull to keep the airstream from blowing in and puncturing the hydrogen cells. The three Navy cruisers and eighteen destroyers that were ready to respond to Rosendahl's call for help would not be needed.

In improving weather, the *Graf Zeppelin* made for the American coast, crossing it just north of Cape Charles, Virginia, the morning of 15 October. When it arrived at Lakehurst later in the day, it had been flying 111 hours and had covered 6,168 miles. Rear Admiral Moffett was on hand to welcome it, only to be greeted by five hundred pounds of water ballast that was dropped directly on him during the ship's landing maneuvers. Sunday's crowd of two hundred thousand had become Monday's of thirty-five thousand.

Sixteen October saw a ticker tape parade up New York City's Lower Broadway to honor Eckener, his officers, and the passengers. The zeppelin man then went to Washington to breakfast with Calvin Coolidge and meet presidential candidate Herbert Hoover. He laid a wreath on the Tomb of the Unknown Soldier in Arlington Cemetery, an unusual honor considering the anti-German feelings smoldering from World War I.

After the Navy put a new cover on its fin, the *Graf Zeppelin* prepared to leave for home. In appreciation of the help given him, Eckener invited the Navy to have three observers on the return flight. They were Cdr. Maurice R. Pierce, Lakehurst's executive officer, Lt. Charles E. Bauch, and Lt. T. G. W. Settle. They boarded the ship carrying winter flight gear. Although promoted as a luxury passenger-carrying aircraft, the *Graf*'s spaces were unheated.

In addition to the three naval officers, there was another special passenger: Willi von Meister.

Friedrich Wilhelm von Meister had been general agent in America for the Maybach company, builder of the engines for the *Los Angeles* and now the *Graf Zeppelin*. Willi sometimes carried a gold watch given him by his godfather, Kaiser Wilhelm II, Emperor of Germany. A bachelor in his late twenties, he stood six feet six inches tall. He was considered extremely handsome by the Broadway stars, singer Helen Morgan for one, whom he liked to date. His English was flawless, reflecting his schooling in Britain. Much of his charm he inherited from his English-born, German-wed, mother. She once so fascinated Winston Churchill that he was prompted to say: "England can ill afford to export women of such quality."

As engineering officer for the *Los Angeles* and responsible for its Maybach engines, Settle had come to know von Meister well. When Willi came to Lakehurst, he stayed with "Tex" and Fay.

The flight back to Germany began with the discovery of a teenage stowaway who was promptly put to work in the galley. (Eckener was not amused, every pound of weight on a lighter-than-air craft being important.) Off Newfoundland, hurricane-force winds and violent turbulence battered the zeppelin so severely that Eckener himself wondered whether the ship would hold together. It did, without failure of a wire, without breakage of a girder, without damage to the outer cover.

The three U.S. Navy observers quickly found the *Graf Zeppelin* had much of what the *Los Angeles* did not. Comfortable staterooms. Gourmet food. Fine wines. Great cocktails. (Not even coffee, the Navy's lifeblood, could be guaranteed on board the *Los Angeles*. A Lakehurst medical officer, deciding too much was being consumed on long flights, directed that soup be substituted. The "L.A." was officially "dry." But it did once return from a flight to the Caribbean with ballast tanks, ones seldom dumped in flight, filled with rum instead of water.)

"Tex" took special interest in how the Germans used Blaugas in the Maybachs. He was impressed by the ease with which the motors were shifted between gasoline and Blaugas as fuel. He did not think Blaugas should be used in U.S. Navy airships, however, not liking to fill a hull with flammable gas. He believed it negated the safety provided by helium. The future of airship propulsion, in his mind, belonged to the diesel engine.

After they arrived at Friedrichshafen on 1 November, von Meister took Settle to the Maybach factory where he met Dr. Maybach. He was intent on perfecting gasoline-fuel engines and had no plans for diesels. But eight years later, Luftschiffbau-Zeppelin's magnificent creation, the *Hindenburg*, would be propelled by diesels manufactured by Daimler-Benz.

Just as the Americans had gone wild with "Zeppelin Fever," the Germans did so also. Many had contributed money to build the *Graf Zeppelin* through the popular subscription effort mounted by Hugo Eckener and his zeppelin captains. Germans considered the *Graf* "their airship" and were intensely and nationalistically proud of it. So was President von Hindenburg, who had Eckener bring the ship to Berlin so he could meet its officers. Pierce, Bauch, and Settle were included in the reception by Germany's president.

When he departed for the States shortly afterward on a German steamship, Settle wondered if, when, and where his path and the *Graf Zeppelin*'s would cross again.

It would be nine months later and in circumstances that almost spelled the famous airship's end.

TEN

1929

T EX SETTLE WANTED TO FLY AIRPLANES IN ADDITION TO AIRSHIPS AND balloons. Admiral Moffett denied his requests for assignment to heavier-than-air training. The BuAer chief wanted to keep this smallish, somewhat dark-complected officer in airships.

Moffett needed him to be a member of the source selection board that would choose the contractor to build the two replacements for the *Shenandoah*. He needed him, also, to be the bureau's inspector of naval aircraft at the contractor's plant.

Congress, in 1926, had approved the building of two new rigid airships for the Navy. They would be the ZRS-4 (Lighter-than-Air, Rigid, Scout, #4), later to be named the USS *Akron,* and ZRS-5, the future USS *Macon*.

The contract to build these rigids, the largest ever up to that time, was awarded in October 1928 to the Goodyear Tire and Rubber Company of Akron, Ohio. The company had purchased the North American rights to the patents of Luftschiffbau-Zeppelin and had put thirteen of its top engineering talent on its payroll. It was an almost impossible combination for any other concern to match, much less better. The ZRS-4 would cost $5,375,000, the ZRS-5 $2,450,000.

Goodyear would construct these "super-dirigibles" in a giant "super hangar" that it called an "air dock" and had erected at its own expense at the edge of Akron's municipal airport. The structure was 1,175 feet long, 325 feet wide, and 211 feet tall. It enclosed 55 million cubic feet of space, far surpassing the Navy hangar at Lakehurst. Ultra-streamlined to minimize air turbulence around it, the dock had rounded, clamshell-type, doors that moved on tracks to the sides when opened.

Here, in this hangar, Settle and his staff—Lt. Roland G. Mayer, Lt. George V. Whittle, and Lt. Cornelius V. S. Knox—had their offices. Their job, and the job of their civilian inspectors, was to oversee the work Goodyear performed. For the purpose of winning the contract and building the ships, Goodyear had created a subsidiary, the Goodyear-Zeppelin Corporation, with the Tire and Rubber Company president, Paul W. Litchfield, also its president. Litchfield, a fervent believer in a tremendous future for large airships, had been the person mainly responsible for the acquisition of Luftschiffbau-Zeppelin patents and technical personnel.

The results of the Navy representatives' labors—and of Goodyear-Zeppelin's—were the sky twins, ZRS-4 and ZRS-5, each 785 feet in length, 133 feet in diameter. In volume, they were 6.8 million cubic feet. When they took to the air, they would be the largest man-made objects that had ever flown. To many who saw them, they were also the most beautiful aircraft ever designed.

Except for weight-saving improvements and streamlining, and for the use of wooden propellers on the first ship and metal ones on the second, they were identical ships. The wooden props gave the ZRS-4 a top speed of seventy-nine miles per hour; the metal props allowed the ZRS-5 eighty-seven miles per hour.

Settle was detached from Lakehurst in January 1929. His orders authorized leave and temporary duty in Washington before proceeding to Akron. He put the leave to immediate use, going to New Brunswick, New Jersey, to take flying lessons. If the Navy wouldn't teach him to pilot an airplane, he would learn to do so on his own. At a civilian flight school, he gathered the necessary flight hours to earn a private pilot's license.

In Washington, he reported to the no-frills office of BuAer's Airship Design Section in the old "temporary" World War I building, with the swinging barroom doors, that housed the Navy on Constitution Avenue. His duty there was to review technical matters concerning the ZRS-4 and -5 and organize plans for the inspector's office he was to run at the Goodyear-Zeppelin plant.

Head of the airship design "desk" was Cdr. Garland Fulton, himself once an inspector of naval aircraft at Friedrichshafen while the *Los Angeles* was under construction. "Froggy" Fulton was quiet, unassuming, and little known. Yet few men influenced Navy Lighter-than-Air as much as he.

In contrast, the civilian engineer who shared Fulton's office, Charles P. Burgess, was talkative and brought humor by the carload to it. A former naval architect, he had made the transition from ships to airships. Author of the only textbook on the design of rigid airships, "C. P." would leave a legacy of design memoranda, analyzing the airship's engineering past and forecasting its technical future.

Settle's duty in the bureau lasted until summer.

On 4 March, it included taking an emergency phone call from the commanding officer of the Naval Air Station, Anacostia, Washington, D.C. The inaugural parade for newly elected President Herbert Hoover had taken place that day. It had included a lighter-than-air flyby of Navy and Army airships led by the *Los Angeles*. The weather had been miserable with rain, low visibility, and a ceiling of a few hundred feet.

After their part in the parade, Rosendahl, who served as grand marshal on board the *Los Angeles,* released the airships. The Army's TC-5 and TC-10 nonrigids hurried home to Langley Field. The "L.A." returned to Lakehurst, but the Navy blimps J-3 and J-4 were stymied by worsening conditions. They landed to wait out improved weather at Anacostia where, unfortunately, there were no mooring masts and no one was on hand who knew how to handle blimps. The Anacostia commander had foreseen a possible problem and asked BuAer to name an experienced airshipman he could call upon if needed. Settle was the man the bureau volunteered.

Dressed in his "blues"—to wear at the inaugural parade—and sporting a brand new cap, Settle went at once to Anacostia, where he found the two airships wallowing in a sea of mud. Disregarding uniform and cap, he waded in to wrestle with them as the winds increased.

The sailors, who had been made available for ground handling from Anacostia, and the ones brought in from other naval activities in the area, had great difficulty keeping their feet and standing fast in the mud. Some were so exhausted by their efforts that they were ordered to the air station's sick bay. Trying to tame the J-3 and J-4 was dangerous business. As the ships were driven up and down by the wind, the men struggling with them risked being caught and pinned underneath.

After the gusts peaked at fifty miles per hour and after the airships' envelopes had been punctured and their cars battered by pounding against the ground, Settle ordered them ripped. They were not lost, however. Shipped to Lakehurst, they would be fitted with replacement envelopes and restored to service.

The Settles were still in Washington, living in a furnished rental apartment on Connecticut Avenue, when the next National Balloon Race was held. Admiral Moffett not only approved "Tex's" participation but also encouraged it. He knew that a win by a Navy balloonist would be helpful to the Navy's image.

The competition would be held on 4 May 1929 from the University of Pittsburgh stadium. Van Orman, his ankle now mended, was there, anxious for another try. Working with electrical insulation expert Arthur Osten and the Ohio Insulation Company, he had a device he hoped would protect him from another lightning strike. It was a cagelike arrangement of wires draped over the basket.

The Army's Lieutenant Ent, his balloon having been set afire and his fellow occupant killed the preceding year, was, like Van Orman, a center of interest. How did he hope to avoid another lightning encounter? "By depending on my rabbit's foot," was his reply.

Settle was more practical:

> I'm depending on the way I fly my balloon. Think of it, particularly when it's wet, as a flying electrical conductor that you want to keep as short as possible when around lightning. To keep it short, you pull in your drag rope, your running light, and your radio antenna, in fact everything hanging below the basket.
>
> You don't valve any gas or pour any sand. Hydrogen rising from the balloon or a stream of sand ballast dropping from it extends the vertical length of the conductor. If you've absolutely got to valve, do it in short bursts. If you have to drop sand, don't do it in a long stream.

Twelve balloons left Pittsburgh, beginning at five o'clock that evening. All were thirty-five thousand cubic feet in size. Two were Navy, three were Army, and seven were civilian.

It was 5:45 PM when Settle, in Navy balloon A-8278, with a new aide, Ens. Wilfred Bushnell, ascended in a light drizzle, climbed to twenty-two hundred feet, and began drifting toward the northeast. At 10:15 AM on 6 May, they were over the Canadian province of New Brunswick, approaching the Northumberland Strait. Beyond it lay Prince Edward Island and, beyond it, the open Atlantic Ocean.

With the mind-set common to all racing balloonists—make as many miles as possible—"Tex" took the balloon across the strait, climbing to ten thousand feet. Nearing the island's shore, he valved down to fifty or a hundred feet off the water, ten miles off its north coast. Dropping the drag rope, he trailed it across the water while drifting at about twenty-five knots, toward the east southeast. They paralleled the island's northern coast and then crossed it at its eastern end. (The drag rope was a kind of automatic pilot that controlled altitude. It was dragged across the ground or water. If the balloon rose, it would pick up some of the rope, become heavier, and descend. If it descended, it would deposit rope on the ground, making itself lighter. Down! Up! Without having to valve or drop ballast.)

Settle and Bushnell landed at 1:05 PM between a house and barn that belonged to Mr. John McAdam of Conway, East Savage Harbor. Farmer McAdam and family had never seen a man-carrying balloon. Frightened by it as it approached, they ran into the house. Not until the bag was lying lifeless on the ground did they reappear.

The Navy's entry had flown 952 miles in forty-three hours and twenty minutes. After examining the flight log and the sealed barograph, which recorded altitude versus time throughout the flight, the NAA declared Settle and Bushnell winners of the race. They had flown twice as far as runner-up Van Orman.

The flight had been a very long one for a thirty-five-thousand-cubic-foot balloon. The FAI confirmed this when it advised Settle he had established a new world's distance record for what it termed Class A (Spherical Balloons), Fifth Category.

The other Navy entry in the "nationals" (Lt. J. C. Richardson and Lt. M. M. Bradley) had not done nearly as well. It landed at Apollo, Pennsylvania, only twenty-three miles from the starting point.

The national race having been flown and won, it was time for Settle to relocate and begin his duties in Akron. The city was far from being an ideal place to live. Two hundred thousand people inhabited it and fifty-five thousand of them worked in its rubber companies. Soot from soft coal was everywhere, as was a pervading stench of rubber. The desirable part of town was upwind, and the Settles were fortunate to find a house in that area.

The Navy people and Goodyear's worked well together but their families seldom socialized. The notable exceptions were Settle, by nature a gregarious and outgoing sort, and Van Orman, who was of a similar disposition. They were rivals in the air and close friends on the ground.

Settle had scarcely arrived in Akron when he was on his way to Los Angeles. Moffett wanted him to take charge of ground handling the *Graf Zeppelin* when it arrived there from Tokyo on a round-the-world flight. The Navy had, in fact, provided a stub mast for the German ship at Mines Field, today's Los Angeles International Airport.

There was a feeling of kinship between the zeppelin operators and U.S. Navy airshipmen. The Navy had volunteered to supply the mast and direct a ground crew at Mines. Eckener had reciprocated by inviting two naval officers—they were Lt. Cdr. Charles E. Rosendahl and Lt. Jack C. Richardson—to make the world flight with him.

In Los Angeles, Settle found preparations to land the airship in almost total disarray. The California National Guard was supposed to have provided the ground crew but the governor had refused to pay the bill. So had the City of Los Angeles.

The L.A. Chamber of Commerce contacted the commandant of the Eleventh Naval District in San Diego to ask if Navy personnel could be made available. The commandant met with Settle and with representatives of the city and of its chamber of commerce. Together, they reached an agreement whereby Navy and Marine Corps personnel would be made available and the city would cover the cost. Meanwhile, the *Graf Zeppelin* was at Tokyo, preparing to take off.

Three hundred seventy-five officers and men went from San Diego to Los Angeles by special train. On Saturday, 24 August, they pitched their tents on Mines Field. The next day, Settle and Karl Lange, a Goodyear blimp pilot and lieutenant (later rear admiral) in the Naval Reserve, put them through their paces, showing them how to line up to meet the ship and take its lines and rails in hand.

On the twenty-sixth, the *Graf Zeppelin* arrived at dawn. It found a warm welcome. It also found a temperature inversion. The air near the ground was colder than the air above it. It was the kind of meteorological condition in which no lighter-than-air vehicle wants to come down. Balloons and airships descending from warmer into cooler air find their buoyancy increased and their landings difficult. The *Graf Zeppelin* was no exception. Eckener had to let off hydrogen to land.

All day on the twenty-sixth, the airship rode to the mast, while hydrogen, gasoline, Blaugas, oil, water, and provisions were loaded. That evening, newspaper publisher William Randolph Hearst, a sponsor of the flight (in exchange for exclusive reporting rights on board), held a banquet in honor of the dirigible at the Biltmore Hotel. Following the dinner, everyone headed for Mines Field to see it leave.

Another inversion lay over it. This time, as the airship tried to rise, it would encounter warmer air and lose lift. Seven crew members were taken off to lighten the load and proceed to the next stop, Lakehurst, by plane. Eckener would have to depend on help from aerodynamic lift, generated by the hull as it gathered speed through the air, to get off.

Willi von Meister described what happened in a letter to his mother:

> Eckener had been ill with food poisoning acquired in Japan and was worried about the trip over the mountains. He wanted to take off at once and was

forced to make a running or dynamic start. That means the motors run at full speed while the ground crew holds on. Then, at command, they push the ship up in the air with all their might. They did so. The ship rose and moved forward with relative ease, picking up speed at a terrific rate, but with her tail down and the rudder plowing a deep rut in the soft and sandy field. This for about 1,000 feet with a fence and high tension electric wires ahead. It looked like she was going to hit them. All of us who knew what was going on were completely paralyzed. The crowd of many thousands must have felt it too. Those seconds seemed like hours. I saw in my mind only those high tension wires and hydrogen gas and countless automobiles and people on the other side of those wires. And the passengers and crew.

It was ghastly when suddenly there was a splash of water ballast forward. The elevators were put hard down which brought the tail whizzing up. The ship cleared the wires by 25 feet. You could see the reflection of the red lights on top of the wire poles against its silvery sides. The fifth engine, which could not be started before the tail was clear of obstructions, started with a roar. Picking up speed, she disappeared into the night.

The crowds were quite silent at the start for they must have realized the danger. But when she cleared the wires, there was such a noise of applause and horns as you have never heard before. I couldn't help it but I cried like a baby out of joy and happiness.

The bottom of the lower rudder had been crumpled but its fabric had not been torn. There was no need for the ship to turn back. It continued to Lakehurst where it landed on 29 August, having circled the earth in twenty-one days total elapsed time, twelve days flying time.

For his performance at Mines Field, T. G. W. Settle received a letter of commendation from the secretary of the Navy, which included these words: "On the departure of the Zeppelin, Lieutenant Settle's plan and advice in connection with casting loose from the mooring mast and walking back the Zeppelin therefrom, was not accepted by Dr. Eckener and, as a consequence, the *Graf Zeppelin* narrowly escaped disaster through contact with high tension wires around Mines Field."

September 1929 brought a new type of airship to Lakehurst. It also brought the opportunity for Settle to fly in the Gordon Bennett.

The airship, designed by Ralph H. Upson and built by the Aircraft Development Corporation of Detroit, differed from previous airships in that it had a metal skin. Identified by the Navy as the ZMC-2 (Lighter-than-Air, Metalclad,

#2), it had no official name but was known by various nicknames including "The Tin Blimp."

Upson, Goodyear's first aeronautical engineer, had been building balloons and airships since before World War I. Winner of one Gordon Bennett race (1913) and three national balloon races (1913, 1919, 1921) he had taught Van Orman how to pilot a balloon. The ZMC-2 was his attempt to overcome the deficiencies he saw in fabric-covered rigid airships. An extremely talented engineer, he would become first to calculate "the natural balloon shape"—it turned out to be the shape of a pear—but not before being honored by the aviation community for his contributions to airplane wing design.

The ZMC-2 depended for its construction upon a "sewing machine" that was the invention of Edward J. Hill. It sewed together the Alclad aluminum sheets that made up the airship's skin. It simultaneously accepted three strands of .035 inch wire, cut the wire to rivet length, pushed it through the sheets of aluminum (.0095 inches thick), and headed it up into rivets. When he wasn't working on or with his machine, "Eddie" Hill was racing balloons. He and A. G. Schlosser had won the Gordon Bennett at Detroit in 1927.

The metalclad ZMC-2 embodied a promising technical approach that was never pursued. *U.S. Navy Photo*

The ZMC-2 looked like an egg and was just about as stable. Although it had eight stabilizing fins, the ship rolled fearsomely, making even the most experienced of airshipmen queasy.

The metalclad was too small and its performance too limited for it to have been anything but an experiment to determine the feasibility of its type of construction. The design never caught on, the Navy preferring more conventional types of airships. Dismantled in 1941 after 752 flights and 2,265 hours in the air, its design was a might-have-been of naval aeronautics. When delivered to Lakehurst from Detroit, the pilot had been William E. Kepner, captain, U.S. Army, winner of the 1928 Gordon Bennett meet. Army and Navy pilots, Kepner being one, often were trained in and flew each other's airships.

The next James Gordon Bennett International Balloon Race was held in St. Louis on 28 September 1929. Captain Kepner and Lieutenant Eareckson, who had won the Gordon Bennett the previous year, were present to defend their championship. Settle was, too, having won the national race four months before. And so, as might be expected, was Van Orman.

There were entries from Argentina, Belgium (DeMuyter), Denmark, France, Germany, and the United States. Coal gas filled the bags, which were eighty thousand cubic feet to compensate by volume for the low lifting power of the gas. The contestants spoke in a multitude of tongues about the St. Louis heat and the fact that the tree-studded grounds of the Laclede Gas Light Company was no place for a balloon meet.

The nine balloons began taking off, one at a time, at four o'clock in the afternoon. The Navy one held Settle and Bushnell, recently promoted to lieutenant (junior grade).

While ascending, a balloon had to be able to vent gas as it expanded with decrease in atmospheric pressure. Otherwise the bag would burst. Gas was expelled through a hole in the bottom of the bag, the same through which the gas was introduced during inflation. A fabric sleeve called an appendix hung from the hole to within reach of the crew. They kept it open to allow excess gas to "valve" itself off as they rose. They twisted and tied it closed to prevent air from entering the envelope as they came down—the combination of air and flammable gas being bad news indeed.

A frontal line of thunderstorms threatened the racers. Some, including Settle, tried overflying them. He and Bushnell went to twenty-six thousand feet without oxygen. They passed directly over a storm cell that thundered at them as they passed. "The physical and mental effort to stay awake and carry on at that altitude," Settle said later, "was enormous."

Van Orman (United States), DeMuyter (Belgium), and Settle (United States), the "winningest" balloon racers of all time. *Goodyear Tire and Rubber Co. Collection, University of Akron Archives*

Van Orman and aide Allen MacCracken had oxygen and reached thirty thousand feet. But to little avail.

The weather won out. By late afternoon on the day following takeoff, all nine balloons were down. Van Orman took first place with 341 miles. Settle was third with 315; he and "Bush" landed in a cornfield near Eaton, Ohio, with one bag of sand and no disposable equipment remaining.

The distances covered had been the shortest in the eighteen-year history of the Gordon Bennett races.

ELEVEN

AKRON AND MACON

IN 1926 CONGRESS APPROVED A FIVE-YEAR PLAN FOR BUAER THAT included two rigid airships.

Goodyear-Zeppelin Corporation, a subsidiary of Goodyear Tire and Rubber, won the design competition and contract to build them. It had acquired the patent rights and had exchanged stock with the Zeppelin Company in Germany. It had also acquired thirteen of its key technical personnel, including Dr. Karl Arnstein, its chief designer.

Arnstein—short, bald, bespectacled, and soft-spoken—was an innovator who had been restrained by the conservatism of the Zeppelin Company. Now, as Goodyear-Zeppelin's vice president for engineering, he had a free hand to pursue his ideas. He abandoned the traditional, triangular zeppelin girders, substituting four-sided box-type ones. He made the frames of the hull deeper and more massive, eliminating a lot of bracing wire by so doing.

His design called for three keels, one along the top and one along each side, about 45 degrees up. Along these lateral keels, he placed the engines, four to a side. They were 560-horsepower German-built Maybachs. Because helium was the lifting gas being used, they could be carried safely inboard. Drive shafts connected them to propellers that could be swiveled to provide vectored thrust.

With no keel along the bottom, there was room to include an airplane hangar, seventy-five feet by sixty feet, in the underside a third of the way aft. It would house five "hook-on" fighters that would leave the ship and return to it on a trapeze that was lowered underneath.

Gas cells would use a gelatin/latex compound instead of intestines from cattle to seal them against leakage.

Swiveling propellers gave the *Akron* and *Macon* vectored thrust. Here, Karl Arnstein, Goodyear-Zeppelin's head of engineering, and Tex Settle examine a test installation. *T. G. W. Settle Collection*

And then there were the fins....

The ZRS-4 and ZRS-5 had fatter hulls—a lower fineness ratio—than had the *Shenandoah* or *Los Angeles*. Their bulkier shape, however, prevented those on the "bridge" in the control car from seeing the lower fin.

Rosendahl, the ZRS-4's prospective commanding officer, remembered that night at Mines Field when he was in the control room of the *Graf Zeppelin* while it dug its fin into the dirt. He insisted he had to be able to see the bottom of his ship's lower stabilizer when on or near the ground. BuAer honored what seemed to be a reasonable request by issuing a change order. It moved the control car slightly aft and deepened the fins.

Goodyear-Zeppelin had originally proposed relatively small, tapered, and streamlined fins, much like those carried on the *Los Angeles* and *Graf Zeppelin*. The Germans had built their stabilizers as parts of the hull itself. One of Dr. Arnstein's innovations was to build them separately and then bolt them to the frame. He planned to use three of the ship's main rings as attachment points.

But the bureau's change made this impossible. In deepening the fins, they had also to be shortened in their fore-to-aft length. Their leading edges no longer reached to a main frame. These edges were unsupported except for being attached, mainly by wiring, to one of the weaker secondaries.

When Hugo Eckener, in Germany, learned of this, he asked his American representative, Willi von Meister, to arrange a meeting for him with Arnstein during his next visit to the United States The meeting was held. Eckener expressed his misgivings about the fin redesign. Arnstein, reflecting the policy of his company president, Paul W. Litchfield, which was to give the customer what he wanted, replied that Goodyear would build for the Navy whatever it desired.

And what BuAer desired were deeper fins!

When the ZRS-4 went down in the Atlantic in 1933, it was for reasons that may have had little or nothing to do with its fins. No one could be sure, however; the loss of seventy-three of the seventy-six men on board made it difficult to determine exactly what happened. When its sister, the ZRS-5, crashed into the Pacific two years later, it was because part of its hull structure, which supported its upper fin, had collapsed.

In May 1930, when construction of the first of the new rigids was half-complete, Settle volunteered for a special flight. He was to drop in a glider from the *Los Angeles* over Washington, D.C.

The U.S. Navy had only one glider, a German-manufactured Prüfling. Lt. Ralph S. Barnaby, later America's foremost gliding and soaring expert, had ridden it down from the "L.A." at Lakehurst in January. The feat was scheduled to be repeated on May Day as part of the Navy's program in the Curtiss Marine Trophy Race at the Anacostia naval air station.

Barnaby was to have been the pilot but fell ill with the flu. There was no planned backup for him. Hearing this, Settle stepped forward. In addition to getting time in flying private airplanes at the Akron airport, he had been taking

Goodyear-Zeppelin's "airdock" where the ZRS-4 and -5 were built. *Author's Collection*

instruction in gliding as a member of the Akron Gliding Club. His instructor was Dr. Wolfgang Klemperer, recipient of one of the first glider licenses issued in Germany, and now Goodyear-Zeppelin's director of research. "Tex" practiced from a hillside overlooking the airport and had quickly qualified for a license.

The flight of the *Los Angeles* from Lakehurst to Anacostia, with the Prüfling attached to its underbelly, encountered unusually turbulent air. The airship was thrown violently around, shaking the glider and shifting it in its mounts. Settle looked on and later admitted to wondering whether the motorless craft would make it to Washington in one piece.

On 31 May 1930, with the *Los Angeles* hove to eighteen hundred feet above the crowd, Settle, who had never flown a Prüfling before, climbed down from the hull to board it. As he did, he saw that the leading edge of its port wing was crushed as far back as the wing spar and that there were holes, one two inches by four inches, in the top of the wing. At 2:15 PM, he cast loose.

The Prüfling fell free and entered into an easy dive. As it did, its pilot discovered the airspeed indicator wasn't working. Settle flew "figure-eights" on the way down, came about into the wind, and landed on the air station at 2:22 PM.

The drop-away over Anacostia was the second and last time a glider was launched from a Navy rigid airship.

A month later, 4 July 1930, "Tex" was aloft again, this time in the 1930 National Balloon Race with Lt. Roland G. Mayer as aide. They lifted off from Bellaire Speedway, Houston, Texas, in a thirty-five-thousand-cubic-foot hydrogen balloon, landing forty-two hours later on a deserted farm near Dover, Tennessee. They placed third with 650 miles. (The 1930 "nationals" included two other Navy balloon teams: Lt. (jg) Wilfred Bushnell with Lt. J. A. Greenwald, and Lt. R. R. Dennett with Lt. C. F. Miller.)

A Goodyear balloon, crewed by Roland J. Blair, pilot, and Frank A. Trotter, aide, won the race. Their distance was 850 miles. (Van Orman, who was automatically entitled to fly in the next Gordon Bennett, having been the winner in the preceding one, had stepped aside to give his colleagues a chance.)

Roland G. Mayer, Settle's aide for the 1930 national race, had been a project engineer for the *Shenandoah* during its construction. Few knew it as well as he did. He would be assigned to duty on board as repair and damage control officer. As a member of the Navy's Construction Corps, he was ineligible to command, but he was qualified as a naval aviator in lighter-than-air and heavier-than-air craft.

Mayer was a "silent airshipman" who was little known outside lighter-than-air circles. Holder of a degree in mechanical engineering from the University of Washington, he had entered the Navy as a reservist in World War I.

He had been on the *Shenandoah* when it broke away from the Lakehurst mast. He had gone forward to the bow to tie down the gas cells that had been ripped open and deflated. He continued to serve on the airship after it was repaired and was one of its twenty-nine survivors when it broke up over Ohio in 1925. With six others, he free-ballooned its forward section for fifty-three minutes before it was brought to earth. (Rosendahl was the senior man among them.) During this ordeal, Mayer rescued the aerological officer, who was barely hanging onto the wreckage directly above a hole and empty space. Mayer threw him ropes and hauled him safely on board.

When his tour in Settle's office was completed, Mayer would be assigned to the ZRS-4 and -5. He was fortunate enough to be transferred from both before they were lost.

In the race out of Houston, Settle and Mayer soon found themselves amid cumulus clouds that were building up. They were fascinated by how they developed and dissipated, while being careful not to be sucked up in one of them.

As they watched, a cumulus would boil up like a geyser. The tops of some would fall over and descend like a waterfall, evaporating on the way down. Others would halt their ascent and sink back, evaporating or thinning as they did. Still others would rise up in a high column and spread out anvil-like on the top; then the column halfway down would lose its energy and evaporate, leaving the top detached from its base.

Settle and Mayer enjoyed their ringside seat to these atmospheric processes. Only one cumulus thundered and threw lightning at them. They got out of its way by changing altitude.

Back in Akron, they began to prepare for the big day: the christening of the ZRS-4. The nation was deep into the Great Depression. Akron, an industrial city, had been hard hit. Something was needed to heighten its spirit. The naming of the ZRS-4 was expected to be it.

The airship's sponsor was the first lady, Mrs. Herbert Hoover. In a gala ceremony, she rose in the air dock to say: "I christen thee *Akron!*" Then she pulled a line that opened a hatch in the aircraft's bow. Forty-eight homing pigeons, one for each state, fluttered out and disappeared through the open hangar door. A band played "The Star Spangled Banner." The *Akron*'s restraining lines were relaxed just enough to let it float briefly a few inches off the floor.

The new airship was undocked for its first flight on 23 September. Onlookers marveled at its size and the majesty of its lines. On board for a cruise over the Ohio countryside were 113 persons, including the secretary of the Navy.

But the great ship was tarnished in reputation even before it took to the air. It was suspected of having been sabotaged. A Goodyear-Zeppelin riveter, one of eight hundred company employees building the ship, was a Hungarian who was heard to brag about his role in the communist revolt led by Bela Kuhn in Budapest in 1919. It was alleged that he would spit on the rivets, as he inserted them into the girders, to induce corrosion. The Office of Naval Intelligence, the Federal Bureau of Investigation, and Settle's office all investigated whether charges should be brought against him. No damage could be found that he had caused. The matter was dropped.

Critics, remembering the loss of the *Shenandoah* and, more recently, the fiery crash of Britain's R-101 in 1930, were quick to take aim at the Navy's new skyship. They particularly disliked the fact that its engines, the eight Maybachs, were purchased from Germany and not made in America. The reason was that America had no aircraft engine that could run for the many continuous hours required by a long-range, long-endurance, scouting airship. American aircraft engines were designed for airplanes, Maybachs for airship use.

After its commissioning, the *Akron,* based at Lakehurst, made an aerial tour of East Coast cities, accompanied by the *Los Angeles;* it was the only time two rigid airships were ever seen in American skies at the same time.

Charles E. Rosendahl was the *Akron*'s first captain, followed by Alger H. Dresel, who relieved Rosendahl so he could go to sea. Dresel was followed by Frank C. McCord so that Dresel could take command of the ZRS-5.

Dresel and McCord were full commanders; Rosendahl was a lieutenant commander, as were Herbert V. Wiley, the *Akron*'s executive officer, and "Tex" Settle. The officers who had the longest experience in balloons and airships were passed over for command of the new rigids, except for Rosendahl.

Moffett was responsible for this. He wanted Lighter-than-Air to have more clout, to "draw more water" as he put it. The two ZRS airships should be captained by commanders, not lieutenant commanders. He selected Dresel and McCord accordingly.

The years 1931 and 1932 were turning out to be good years for Navy airshipmen: The *Akron* was flying. The ZRS-5 would soon be named the *Macon*. The *Los Angeles* was still in service, although scheduled for decommissioning and layup. A "West Coast Lakehurst" was in the works. And a new developmental nonrigid, the K-1, had been delivered.

The K-1 came into being via budgetary subterfuge. The Navy had only two blimps at the time, the J-3 and J-4. (It also had the metalclad ZMC-2, but its internal rings supporting its aluminum skin hardly qualified it as a nonrigid.) The "J's," looking older than they were, with their open-air cars hanging by wires from their hulls, left a lot to be desired. Airshipmen wanted something new and better. They wanted, also, a blimp they could experiment with. Among other things, they wanted to evaluate Blaugas as a substitute fuel for gasoline.

According to the rules of the game, as it was played in those days, only Congress could authorize a new airship, regardless of type or size. Given the state of the economy, there wasn't a chance it would vote the money. So someone came up with the idea to pay for it out of maintenance funds and build it in two parts. Goodyear would manufacture the envelope, and the Naval Aircraft Factory, the car. Put them together and you'd have a new nonrigid that you could call the K-1.

The Navy's latest blimp first flew in the fall of 1931. It was 219 feet in length and had a volume of 320,000 cubic feet, of which one-sixth was for Blaugas. It incorporated various innovations—having its car enclosed and mounted flush against the bottom of the envelope, for example. But, if ever there was an "ugly duckling" airship, the K-1 was it. Blessed with a pronounced tail-droop, it was an awkward-looking craft on the ground and in the air. Nevertheless, it would remain in service until World War II. The gaseous fuel tests conducted with it convinced the Navy that Blaugas was not what it wanted.

Moffett Field, transferred to the Army in 1935 and returned to the Navy in 1942. The *Macon* is on the field. *U.S. Navy Photo*

As for the West Coast Lakehurst, it was being built at Sunnyvale, Mountain View, California, down the peninsula from San Francisco. When completed, it would have a hangar 1,117 feet long, of like design to the Goodyear-Zeppelin air dock in Akron. The most modern airship base yet built, it was also, thanks to the Spanish architecture of its buildings, the most attractive.

The ZRS-5, as the USS *Macon,* would begin operating there in October 1933.

TWELVE

The Big Win

In 1931, Settle and Bushnell won the national race. They also won the follow-on international meet.

Van Orman was the victor in the 1930 Gordon Bennett race from Cleveland, landing with Allen MacCracken outside Boston. This win by an American team meant that the United States would be the host country in 1931. Owing to the worldwide economic depression, foreign balloonists could not afford the trip to the States. The "big bag classic" was postponed until 25 September 1932 and moved to Europe—to Basel, Switzerland.

The national race in America was held in 1931 just the same. It started from Akron. The day before takeoff, the pilots of the six balloons that had been entered met at Akron's airport. Each reached into a loving cup to draw out an envelope. Inside was a number, the order of precedence in taking off. The U.S. Army, with two balloons, drew the first and last positions. The Navy drew the fourth position.

At 5:25 PM, 19 July 1931, Settle called out "Let go!" He and "Bush" were off.

There had been the usual preflight hoopla, which included presentation of flowers to the crews by pretty local girls. Settle took the spray and tucked it into the rigging above the basket. Many balloonists considered flowers bad luck and would get rid of them at the first opportunity. Not Settle! He had no superstitions.

Also, at the last minute, there were kisses from the wives. There was none for Bushnell. The twenty-eight-year-old Naval Academy (Class of 1926) graduate was a bachelor.

The Navy takes off! Bushnell and Settle (left to right) watch the ground drop silently away.
T. G. W. Settle Collection

Natural gas, which could only lift twenty-five pounds per thousand cubic feet, was used to inflate the balloons. (A thousand cubic feet of hydrogen could lift sixty-eight pounds, a thousand cubic feet of helium sixty-two.) The entries were all eighty-thousand-cubic-foot bags. Despite the large volume, the Settle-Bushnell balloon carried just twelve bags of sand as it left the ground.

Difficulties began even before takeoff. Turbulence from the wind flowing over and around the air dock, streamlined though it was, made the entries difficult to manage. With little lift to support them and hold them erect, they spun around, leaning way over and even threatening to roll their envelopes on the ground. Navy A-8476 wasn't "walked" to its release point, it was towed there by its drag rope. Its two air-inflated pontoons, Van Orman's invention, were pounded into the ground. One was deflated, the other torn partway off. The rain shield, a waterproof cover over the top of the basket, was punctured. In this fine condition, the Navy team went to five thousand feet to begin the drift northeastward.

The Weather Bureau's forecast of thunderstorms proved accurate. The task at hand was avoiding them as they loomed all around. So as not to overhaul those ahead, Settle descended, put over his drag rope, and hung back while the storm cells distanced themselves from him.

By half past seven, the balloon was surrounded by lightning flashes. Alternately dropping ballast and valving, Settle was trying to stay in a hole he found between them. The contest between Navy balloon and weather lasted through the night. Over mountainous countryside in Pennsylvania, one of the storms passed directly overhead while Settle kept as low as he could, dragging the basket through the tree tops. Rain fell in torrents.

Shortly after four o'clock in the morning, he had to call it quits. They were at fifty-two hundred feet. A heavy thunderstorm was approaching them from astern, and they had only three bags of ballast remaining. Settle wanted to stay ahead of the nearing storm as long as possible and then land rapidly in front of it. He started his descent at 5:00 AM and at 6:00 AM landed and ripped the balloon. It was on a farm at Marilla, New York, eighteen miles east southeast of Buffalo. The envelope was blown over by a strong cool wind out of the west. There was heavy rain. Lightning struck and killed a cow in the adjoining field.

Mrs. Grosendahl, on whose property they had landed, and her neighbors quickly showed up to offer help. Bag and basket, together weighing about eight hundred pounds, were loaded onto a truck for transport to East Aurora and return via Railway Express to Akron.

"What do thunderstorms look like from above?" the express agent wanted to know.

"We didn't have a chance to see them from above," Settle told him. "We were too low, not over seven thousand feet. You have to be much higher than that to be above an electrical storm. When you are, it is a beautiful sight. The tops of the clouds are boiling and light up just like a frosted incandescent bulb at home, flashing on and off. You can see, also, the lightning from the inter-cloud discharges and the cloud-to-earth discharges."

Bushnell joined in the conversation to add a truism for balloon racing: "Whenever there is a race, a thunderstorm convention follows along."

"Tex" and "Bush" returned to Akron, Settle to his duties as inspector of naval aircraft, Bushnell to his assignment as an officer of the ZRS-4 (*Akron*), which was soon to make its initial flight.

The next Gordon Bennett, as mentioned, was changed to 1932 and to Basel. The bags were to fly with eighty thousand cubic feet of coal gas. Settle planned to take "tried and true" Navy A-8476 to Basel. Moffett was unable, in those money-strapped days, to make the funds available. Settle went personally to Admiral Pratt, the chief of naval operations, and from him was able to obtain some partial financial support. Settle and Bushnell would make up the difference from their own pockets.

In Basel, they found the Swiss had outdone themselves preparing for the meet. Each participating team was given its own concrete pad and metal shed, protected by a sentry. Each site had a telephone. Swiss "balloon troops" provided the manpower to unfold, spread out, inflate, and rig each bag for flight. Settle thought it the best-organized balloon race he had ever seen.

However, just two days before the start, he was advised that the documents entering the Navy's balloon, also Goodyear's, were missing. Some means had to be quickly found to certify the volume of the bag. Fortunately, the race committee was satisfied when the U.S. consul in Basel affixed his seal to a blueprint of A-8476 that Settle had with him. This, plus payment for the second time of the twenty-dollar-per-balloon entry fee, got the Navy into the race. Van Orman entered the race in the same way.

The Navy and Goodyear balloons were the only American entries. Their crews carried an international ballooning passport that identified them as race participants and requested safe passage for them in whatever country they might land.

Sixteen teams, buoyed by coal gas, took off on 25 September 1932 from Basel's new gas works. The Navy was twelfth into the air. Settle had the advantage of possessing weather information not available to the other pilots. He had arranged with Dr. Eckener for Luftschiffbau-Zeppelin to make upper air soundings at Friedrichshafen and telegraph them to Basel. There he shared this

Basel, Switzerland, starting point for the 1932 Gordon Bennett. The Navy team won by flying to Daugieliszki, Poland. *T. G. W. Settle Collection*

information with Van Orman and the pilots of the three German balloons that were competing.

The Navy flight began at 4:40 PM with forty-seven sandbags on board. It drifted across Germany and Czechoslovakia. It drag-roped across Poland, terrifying horses, cows, and chickens and sending people running for their homes as the Manila line trailed over rooftops and made loud cracking noises as it whipped around houses and posts. About ten o'clock on the morning of the twenty-seventh, they approached a large forested area and had little more than landing ballast remaining. They came down at Daugieliszki, Poland, having flown 963 miles.

The Goodyear balloon had been unable to make it through the second night. The Navy men had enjoyed the good fortune of good weather, but Ward Van Orman and Roland J. Blair were caught and held by a thunderstorm for nine hours, spending much of that time at twenty-four thousand feet whiffing oxygen. In the electrified atmosphere, sparks shot from Van Orman's fingertips.

The Goodyear bag landed in a hayfield near Kaunas, Lithuania, 841 miles from Basel, earning second place behind Settle and Bushnell.

Local farmers—none understood English except one who had lived in Chicago—helped get Navy A-8476 loaded into two horse-drawn wagons. When the wheel of one collapsed, a Polish army unit happened along. It took over the load. With the balloon and basket riding and pilot and aide walking, the little group made its way to the nearest railroad town, Ignalino.

The commander of the local Polish garrison guarding the border with Lithuania, Capt. Marjan Glut, was cordial and helpful. He had his men unload and pack the balloon for shipment by rail to Warsaw. He invited Settle and Bushnell to be his house guests for as long as they were in Ignalino, which turned out to be twenty-four hours. From Ignalino they sent telegrams to the race committee in Basel, also the all-important barograph, with its instrumented record of the flight, and the flight log. They were the winners!

After a triumphant return to Basel, Settle was guest of honor at a luncheon in Paris given by the American Club. Then he and Fay, who had accompanied him to Europe, took the train to Friedrichshafen to board the *Graf Zeppelin* for its next flight to Brazil. Dr. Eckener, out of friendship for the U.S. Navy and for Goodyear, had invited the Settles and Van Orman to fly to South America as his guests.

No one begrudged Settle this treat. In the corridors of BuAer, he was the man of the hour. Joy Bright Hancock, whom the reader has met, was working again in the bureau. "From the Admiral [Moffett] down," she wrote to a friend, "everyone was simply delighted over 'Tex's' performance. The Admiral was like a small boy over the news."

When the Settles boarded the *Graf Zeppelin* on 9 October, it was the first time "Tex" had seen the ship since that night in August 1929 when it had narrowly missed the power lines at Mines Field. Since then, it had explored the Arctic, even landing on the water off Franz Josef Land. In 1930, Eckener had taken it on a proving flight to Rio de Janeiro. By steamship, it took twenty-one days to travel between Germany and Brazil. By zeppelin, it took a little more than three and a half days. A large German population lived in Brazil, wanting faster transportation to their homeland. Providing that was what Eckener had in mind. The Brazil flight of 1930 was followed by three more in 1931 and by nine in 1932. The Settles and Van Orman were on the eighth of the 1932 flights.

The *Graf* could accommodate twenty passengers in ten cabins inside its control car. Frequently it was sold out. On this flight, however, there were only twelve passengers. One was Charles Dollfus, director of France's Musée de l'Air.

He was also a balloonist. Settle had never raced against him but knew his achievements well, as did every other balloonist of the day. Dollfus was an *aéronaute extraordinaire*. When he died in the 1980s, he had made, some said, more gas balloon ascensions than any other person.

The Settles and Van Orman were going to Rio where they would take a steamer home. The *Graf* was going there, too, but was booked solid from Pernambuco (Recife) to Rio.

So, at Pernambuco, the three got off. There they flew in a Junkers W-34 seaplane, operated by the German Servicio Aereo Condor, the rest of the way. With stopovers, it took thirty-four hours. The aircraft was so loaded with freight that its toilet facilities were blocked. At stops, Fay Settle had to be rowed ashore to look for a bathroom. The sight of a white woman—dressed in winter clothes, disembarking from an airplane, and anxiously searching the area—attracted tremendous curiosity. Wherever she went, a crowd of locals followed her.

By the time "Tex," Fay, and "Van" got to New York on a ship of the Furness Prince Line, the *Graf Zeppelin* had returned to Friedrichshafen and even completed its next flight to Brazil.

In New York, a motorcycle escort accompanied the winner and runner-up of the Gordon Bennett to the Roosevelt Hotel and then to the steps of City Hall for a welcome. It was the high-water mark for Navy ballooning.

But not for "Tex" Settle. The following year, he would fly a non-Navy balloon to a world's supreme altitude record and, at the same time, become the first American to pilot a pressurized cabin into the upper air.

THIRTEEN

HIGHEST ALOFT

On the day that the ZRS-5, which had been christened and was now the USS *Macon,* made its first flight, a telegram arrived at the offices of the inspector of naval aircraft, Akron.

It was from the A Century of Progress Exposition, also known as the 1933 Chicago World's Fair, and was addressed to Settle: "Will you participate in the proposed ascension in the stratosphere from Soldier Field Chicago under the sponsorship of the A Century of Progress Exposition? The flight will probably take place the latter part of June or first of July. The Piccards will probably also participate."

The telegram's date was 21 April 1933. It could not have come at a worse time. Just eighteen days earlier, the *Akron* had crashed into the Atlantic off New Jersey. Settle was busy preparing for the board of inquiry. He was also preparing for the *Macon*'s acceptance and trial flights.

He was intrigued. He had long wanted to take a balloon as high as it could go. In the 1920s, he had approached C. P. Burgess in BuAer with an idea for a pressurized cabin in which to ride a balloon into the upper air. Together they had designed a cylinder, six feet long and three feet wide, that was barely big enough to hold Settle, his life support system, instruments, and flight controls. Admiral Moffett had approved it but had to retract because Congress was objecting to the number of unorthodox aviation projects the Navy was undertaking. The capsule was cancelled. (Even before the Settle-Burgess "flying coffin," U.S. balloonists had tried to reach record heights in open baskets. Army Capt. Hawthorne C. Gray attained 42,470 feet in September 1927 but died when his oxygen supply failed him.)

The telegram was the result of Dr. Auguste Piccard's presence in Chicago to seek exposition sponsorship for what would be his third stratosphere flight. In Europe, the Swiss-born physicist had reached 51,000 feet in 1931 and 53,000 feet in 1932. As protection against the rarefied air, low pressure, and cold, he had taken his earthly environment up with him inside a sealed spherical aluminum gondola.

Piccard made his case. He impressed the fair's officials. He also "hammed it up" for the press who saw him as an eccentric scientist who wore a wicker egg basket upside down on his head as a crash helmet.

But Auguste Piccard was not an American. And according to the FAI, any record established would be "homologated"—FAI speak for "ratified"—in the nationality of the pilot. The exposition wanted it to be in the name of the United States. It looked elsewhere for someone to fly the balloon and had decided to offer the opportunity to Settle.

He would have to take leave to do it. He obtained the needed permission personally from the chief of naval operations.

There were three main sponsors: the A Century of Progress Exposition, the *Chicago Daily News,* and the National Broadcasting Company, which planned to make the first broadcast ever from the stratosphere.

Goodyear-Zeppelin designed and built the six-hundred-thousand-cubic-foot envelope for eleven thousand dollars. The Dow Chemical Company fabricated and donated the seven-foot-diameter Dowmetal gondola. Union Carbide and Carbon supplied the hydrogen—as pure, it was said, as a Louisa May Alcott heroine.

The scientific equipment was mostly provided by Dr. Arthur H. Compton of the University of Chicago and Dr. Robert A. Millikan of the California Institute of Technology. Both were Nobel laureates. Both wanted to observe cosmic rays as they entered the top of the atmosphere from space. The flight would offer the opportunity to record them at a higher altitude and for a longer time than before. Much of the gear consisted of counters, chambers filled with a gas that ionized when a ray passed through. The equipment was checked out and installed by Luis W. Alvarez, himself a future Nobel Prize winner.

As a preflight test of airtightness, the gondola shell was filled with water and with air at high pressure. A soapy solution, applied to its outside, would reveal by bubbles the presence of leaks.

In another test, Settle sealed himself inside, with some bottles of beer, to spend the night checking his oxygen supply and the scrubbing system to remove carbon dioxide. Oxygen would be replenished in flight by pouring it out in a liquid form from Thermos bottles.

After Auguste Piccard returned to Europe, eliminated because he was not American, it was assumed his brother, Jean, would still fly as scientific observer. On 23 July came the announcement that he, too, had withdrawn so that, without his weight, the greatest possible height would be reached.

Settle would go it alone!

Soldier Field was on the Lake Michigan shore, adjoining the fairgrounds. Its stadium could seat thousands. Tickets were sold to see the takeoff. As the public looked on, the balloon, called *A Century of Progress,* was inflated in the middle of the field. Fully inflated, the bag, made of single-ply, rubberized cotton, measured 104 feet in diameter. Only at peak altitude would its hydrogen actually fill the bag. It would take off less than a quarter full to allow the gas to expand during the ascent.

The flight plan was to take off at about midnight, spend the rest of the night at "open hatch" altitudes of fifteen thousand feet or less, and then close up and batten down at dawn as the sun began to warm and expand the hydrogen. He would ride the balloon as high as he could, making sure to retain enough ballast for a controlled descent. Then, at sunset, as the gas cooled and contracted, he would come down.

On 4 August 1933, the weather looked good. The bag was laid out on the field. Copper tubing was connected from 750 hydrogen cylinders to a fabric inflation sleeve extending into the balloon.

Inflation began at ten o'clock in the evening and lasted until after one. As it did, a bubble of hydrogen began building in the top of the bag. As more gas entered, the envelope increasingly filled out and stood taller. It was kept earthbound by ropes through eyelets in a fabric band encircling the bag near its top. When inflation was completed, the gondola was wheeled out on a carriage and attached to the rigging.

At 2:15 AM, as the *A Century of Progress* sat waiting, it was time to check the valve. Settle was concerned about it because the partly inflated bag hung in loose folds at its bottom. The valve cord had to pass through these folds. To prevent it from being restrained by them, the cord had been graphited to make it slippery and brought straight down inside and out the bottom. This, it had been thought, would keep it from binding.

After the public address system called for absolute quiet, Settle gave the cord a long, hard pull, and then abruptly let it go. He was listening for the sound of the valve doors slamming shut. Instead he heard a whistling and hissing that gradually faded and stopped.

Van Orman, whom "Tex" had designated alternate pilot, was standing near him. He heard Settle say, "The damned valve isn't closing!"

The *A Century of Progress* is inflated at Soldier Field, Chicago. *T. G. W. Settle Collection*

In his lieutenant commander's uniform and Navy wings, Settle stood on Soldier Field and looked at the crowd. He thought it too dangerous to deflate the balloon under the circumstances. Besides, optimistic as he always was, he thought the valve might clear as he gained altitude. So he took off. At 3:05 AM. In an almost dead calm. Army searchlights followed him as he rose three feet per second and began drifting west southwest.

At five thousand feet, he had cleared the downtown Chicago area and was over railroad yards and vacant lots. He tried the valve again.

This time it showed no signs of closing. Once more he took hold of the line, hauled it down as far as he could, and let go of it. The valve remained open and gas poured out. The balloon began to fall, Settle dropping ballast all the way. Balloon, gondola, occupant, and contents crashed down on the Chicago, Burlington, and Quincy railroad tracks at Fourteenth and Canal Streets. He ripped and deflated the bag. Except for a bump on the forehead, he was unhurt.

Time in the air had been about twenty minutes. A Chicago newspaper headlined the flight "Settle up!" It later headlined it "Settle down!"

A crowd quickly gathered, many in it smoking. There was hydrogen trapped inside the fabric on the ground. Settle tried unsuccessfully to wave them away. The Chicago policemen who showed up were of little help.

At the stadium, Maj. Chester L. "Mike" Fordney, seeing the balloon descend, jumped into a car, taking along four other Marines. They arrived as the mob was beginning to cut up pieces of the balloon for souvenirs.

"In the ensuing three minutes," reported the *Chicago Daily News,* "the mob was treated to a gala performance of language and action that have won reputations for potency from the Halls of Montezuma to the Shores of Tripoli." The balloon was rescued, rolled up, and with its gondola, taken to a warehouse to be guarded the rest of the night by Fordney's men.

"Tex" would try again, next time from Akron. It was better suited meteorologically than the "Windy City." It had the empty Goodyear-Zeppelin air dock, recently vacated by the *Macon,* for Settle to inflate his balloon inside. He could keep it there, inflated and ready to be taken outside and released when the winds were favorable. Patched in more than two thousand places, the bag was inflated in the air dock 17 November.

This time, Settle was not going to attempt the flight alone. He had selected an aide, Chester Fordney, the forty-one-year-old reserve Marine major who had saved the *A Century of Progress* from vandalism. Moreover, Fordney was interested in science, something that favored his going along.

Inside the air dock, the balloon stood inflated, 164 feet high. Again its fabric hung in folds. This time, however, the valve cord exited not through the bottom but through a flexible tube high up the side of the bag where it was more filled out. The valve worked satisfactorily.

On 20 November 1933, the massive hangar doors were peeled back. The *A Century of Progress*—its name had not been changed although its takeoff place had been—was moved outside by a ground crew of Naval Reservists and civilians. At 9:30 AM it took to the air, Settle atop the gondola, dumping sand ballast, and Fordney inside. Each wore a light leather jacket over old work clothes. Settle also had on sneakers to prevent slipping as he climbed around outside. The surface wind was out of the northwest at nine miles per hour.

A telegram was sent to the Navy Department: "Stratosphere balloon Lt Cmdr Settle Major Fordney took off Akron naught nine three naught. Please

Settle's balloon, inflated, waits in the Goodyear-Zeppelin hangar. *T. G. W. Settle Collection*

inform OpNav, BuAer, major general commandant." General Ben B. Fuller, the Marine Corps commandant, had more than a passing interest in the flight. Fordney was married to his daughter. Fuller took a rather dim view of the whole thing, particularly after he learned that his son-in-law might have to bail out if Settle ran short of ballast coming down.

For about three hours, the balloon loafed along with hatches open at between two thousand and five thousand feet. At 12:45 PM, the hatches were closed and lead shot dropped continuously as it began to climb skyward in earnest. Beneath was East Liverpool, Ohio.

Settle and Fordney were in a gondola crammed with cosmic ray apparatus, an instrument that measured sun and sky spectra, and two cameras, one equipped with infrared filters. Also on board was a device to observe light polarization, flasks to sample the outside air, and color charts to compare with the color of the sky as it deepened from blue to dark blue as they ascended. To study the possible biological effects of cosmic radiation, plant disease spores hung outside the gondola. Unfortunately, the hoped-for spectacular photos of the earth were not obtained. A thick haze in the lower atmosphere left only a small part of the surface visible and that directly below.

The *A Century of Progress* remained at peak altitude for two hours, from 2:10 PM to 4:15 PM, at which time it started back down. At 26,500 feet, the hatch covers were heaved overboard with small parachutes attached. It was almost dark and the Atlantic coastline was near. Settle brought it gently to a landing in a New Jersey marsh.

Its two occupants spent the night rolled up in the balloon's fabric trying to get warm. Next morning, Fordney stripped to his skin and, holding his clothes above his head, struck out for land. He came to a farmhouse five miles away from which he telephoned the balloon's position—where the Delaware and Cohansey rivers joined.

Police, Navy, and other officials began arriving at the scene. The balloon was rolled up. A Coast Guard airplane landed nearby to take Settle, Fordney, and the balloon's two Friez barographs to Washington. There the National Bureau of Standards broke the seals to examine their altitude-versus-time records. Only one had been working. It showed Settle and Fordney had reached 18,665 meters or 61,237 feet. It was a record height.

Settle would receive many honors. The FAI awarded him its Henri de la Vaulx Prize. He received America's Harmon Trophy for the second time (he had received it the first time for winning the Gordon Bennett race).

A telegram from Maxim Litvinoff, foreign commissar of the Union of Soviet Socialist Republics, reached him at the Army and Navy Club in Washington: "Hearty congratulations on your great achievement. I am sure that your colleagues in the Soviet Union have watched with great interest your flight. May both our countries continue to contest the heights in every sphere of science and technique."

Litvinoff's words were prophetic. The Soviets were challenging the stratosphere with balloons of their own. One had already exceeded the Settle and Fordney altitude two months before (20 September 1933) and by a thousand feet. Soviet balloonists Prokofiev, Gudenoff, and Birnbaun had flown a sealed-cabin

William Enyart of the National Aeronautic Association receives one of the two barographs carried by Settle and Fordney. *T. G. W. Settle Collection*

balloon, the *USSR*, to 62,230 feet. Their achievement was not officially recognized because the USSR was not a member of the FAI.

On 30 January 1934, the *USSR*'s flight was followed by the *Osoaviakhim*'s. Its ball-like gondola broke loose and fell to earth, killing all of its three-man crew. Another balloon, the *USSR-Ibis,* tried for the upper air 26 June 1934, only to have its bag fail during flight. Two of its occupants parachuted to safety, the third managing to bring the lightened craft to a landing. There were no more Soviet manned stratosphere balloon flights after that.

The Settle-Fordney record would stand for two years, until the U.S. Army Air Corps/National Geographic Society balloon, *Explorer II,* carried Capt. Orvil A. Anderson and Capt. Albert W. Stevens to 72,395 feet on 11 November 1935.

FOURTEEN

FINIS FOR THE RIGIDS

It was early 1935.

For twelve years, the Navy had operated rigid airships: the *Shenandoah, Los Angeles, Akron,* and *Macon.* There had been a fifth, the British-built R-38, but it had crashed before the United States was able to take delivery of it.

The *Shenandoah* had been torn apart over Ohio in September 1925.

The *Los Angeles,* decommissioned in June 1932, sat retired in the hangar at Lakehurst.

The *Akron* had been a victim of thunderstorms in April 1933 off the New Jersey coast.

The *Macon* was in its second year of service.

With Rosendahl, its first commander, the *Akron* got off to a good start, demonstrating its airlift capacity by carrying 207 persons, and, as a naval scout, searching for and finding "enemy" units off the East Coast. Its early good luck was followed by bad. In February 1932, it crumpled its lower fin during undocking, while members of Congress looked on. In May, landing to a mast at Camp Kearney, San Diego, it pulled three sailors off the ground with its handling lines, one of them able to hang on, the other two letting go and dropping to their deaths.

Charles E. Rosendahl was succeeded by Alger H. Dresel who was, in turn, succeeded by Frank C. McCord. Under their captaincies, the state of the airship art was brought to new heights by the *Akron* in condensing and recovering water from its exhaust gases (to compensate for the weight of the fuel consumed) and in launching and recovering airplanes in flight. Water recovery had been initiated with the *Shenandoah* and *Los Angeles,* the hooking-on and dropping-off of aircraft with the *Los Angeles.*

With McCord as commanding officer and Herbert V. Wiley as his executive officer, the *Akron* departed Lakehurst at 7:38 PM on 3 April 1933. On board was Admiral Moffett, a frequent visitor at Lakehurst.

The airship soon encountered thunderstorms accompanying a cold front from out of the west. To try to avoid them, McCord took the ship out to sea. About midnight, it went into the Atlantic tail-first and lost seventy-three men. Wiley and two enlisted ratings, the sole survivors, were picked up by a passing German tanker. Moffett was lost, McCord was lost, and so was "Tex" Settle's ballooning companion, Wilfred Bushnell. The *Akron* had not carried life jackets.

What had happened? Was the airship driven into the water by downcurrents from the storm cells? Or did it suffer structural failure? There were no satisfactory answers.

The day following the *Akron*'s crash, Lakehurst also lost the J-3. Sent out to look for survivors, it experienced engine failure and ditched in the surf off shore. Two crew members lost their lives.

The *Macon* had been placed in commission inside the Akron air dock on 23 June 1933 by the new chief of BuAer, Rear Admiral Ernest J. King (during World War II, he would become chief of naval operations and commander in chief of the U.S. Fleet). King was sympathetic to airships and wanted to see them prove their usefulness to the Navy's satisfaction.

On commissioning day, with King in the control car, the *Macon* left for Lakehurst. It would be based temporarily there, "shaken down," and readied for the transcontinental flight to its permanent home at Sunnyvale, California.

Cdr. Alger H. Dresel commanded and handled the ship well but with so much caution some thought it "downright embarrassing." You couldn't much blame him. He was responsible for the Navy's only remaining active rigid airship. Its twin had just crashed, cause unknown, with the heaviest loss of life in aviation history. Despite his care and caution, he would nevertheless face an incident that could have cost him his ship.

It took place during the *Macon*'s flight to the East Coast in April 1934. Its route was across the southwestern United States through and over the highlands of Arizona and Texas. To reach the altitudes required, the *Macon* had to go above the altitudes that its helium alone made possible. It would have to be driven, up by the bow, to generate aerodynamically the additional lift required. Its engines would have to run at full power and uninterruptedly for hours.

Dresel knew this, of course, and had prepared for the turbulence and vertical gusts that lay ahead. He did not anticipate, however, that the rough air over

After the *Akron* crashed, the *Macon* (shown here) was the Navy's only operating rigid airship. *U.S. Navy Photo*

southwest Texas would cause girders of the hull to buckle on the portside near the horizontal fin. The ship was in serious danger! He could not turn around and go back through the same conditions again. He had to keep going. His destination was Opa-locka, Florida, outside Miami. A mast and Navy ground party were there.

The *Macon* was saved by its crew that day, particularly by Chief Boatswain's Mate Robert J. Davis. With wooden two-by-fours that he had suggested be put on board because "they might sometime come in handy," he and others shored up the broken framework. Their makeshift repair was enough to get them to Florida.

Goodyear-Zeppelin personnel rushed to Opa-locka with replacement girders. Nine days later the *Macon* was back in the air. While being repaired, it rode to the mast with its hatches wide open to offset the Florida heat. Families of owls entered through the openings, liked the perching and nesting places they found inside, and moved in. It proved as difficult to get them out as to get the repairs done. Even the ground around the mast was a problem. It was alive with snakes.

From Opa-locka the airship took part in Caribbean maneuvers. It performed poorly in its role as a scout. Throughout its lifetime, the *Macon* would be bedeviled by the assessment of war games umpires that it had been "shot down." Almost always this was because it had been assigned a tactical scouting role, rather than a strategic one, the role for which it was built.

The *Macon* had a cruising range of six thousand miles. It should have been used far forward, out in front of the fleet. Instead, the powers-that-were, who planned the exercises and maneuvers, made it operate in limited geographical areas that abounded in "unfriendly" ships and aircraft.

Fleet commanders were becoming impatient with its performance. Admiral King was impatient, too. He believed the *Macon* could and should do better. Dresel's withdrawal from an exercise to take the ship back to the safety of its hangar calmed these feelings not one bit.

Lt. Cdr. Herbert V. Wiley relieved Cdr. Alger H. Dresel on 11 July 1934. At last, the man who had started his lighter-than-air career on the *Shenandoah* and had probably accumulated more flight hours in rigid airships than any naval officer had his own command.

Before Wiley took over, Admiral King instructed him to demonstrate the *Macon*'s usefulness to the satisfaction of the Navy. "Doc" Wiley set out to do this, using the complement of five small F9C-2 Sparrowhawk fighter planes carried on board. Heretofore they had been visualized as mainly protecting the *Macon* from air attack. Wiley planned to use them as the *Macon*'s eyes.

Pilots of its heavier-than-air unit had become accustomed to Dresel's conservative approach to things. They expected that Wiley, a lighter-than-air "old timer," would be the same.

The new captain wanted the F9C-2s to do the scouting, not the *Macon* itself. They were small aircraft, twenty feet long with a twenty-five-foot wingspan. Their top speed was 176 miles per hour, their operating radius 175 miles. To improve upon that performance, Wiley had their landing gear removed. Wheels were not needed while at sea, landings and takeoffs being via the trapeze. A thirty-gallon belly tank was installed. The changes increased the airplanes' speed to two hundred miles per hour and their flight time to five and a half hours. To enable the Sparrowhawks to scout out of sight of the airship, a low frequency radio homing device was developed.

This was all well and good, but it didn't provide the show-of-usefulness King had demanded. Wiley came to the *Macon* with an idea for a flight that would. He implemented it only seven days after taking command. Knowing it would never be approved by his superiors in the fleet, he simply told them he was taking the ship for a protracted flight to sea.

Herbert V. "Doc" Wiley temporarily commanded the *Los Angeles* and was executive officer of the *Akron* (one of its only three survivors). He was captain of the *Macon* when it failed and fell. *Author's Collection*

Franklin Roosevelt was en route from Panama to Hawaii on the cruiser *Houston*. Wiley's idea was to carry out a search-and-locate mission to find the vacationing president of the United States. Two of the Sparrowhawks did—sixteen hundred miles from shore!

They caused some anxiety on the *Houston,* which had received no notice that they were coming. The *Macon* was not in sight, so there was uncertainty about whose planes these were, what their intentions were, and where they came from. When one of those on board pointed to the hook on the top of their upper wings, the mystery of their origin was solved. Shortly afterward, the dirigible itself came into view.

Curtiss-built F9C-2 Sparrowhawk. *U.S. Navy Photo*

The little planes, sans wheels, put on quite a show. They dropped San Francisco newspapers for the president, also specially franked envelopes for his philatelic collection. Roosevelt was delighted and radioed a "Well done!"

The *Macon* returned to Sunnyvale, its commanding officer elated by the successful demonstration the ship had given before a very important audience. But the elation was short-lived. On the way home, messages were received from the fleet demanding to know the circumstances of what he had done. Admiral Joseph N. Reeves, Commander-in-Chief, U.S. Fleet, would call Wiley's actions "misplaced initiative." But he was not relieved of command.

In Washington, the chief of BuAer, Admiral King, was delighted.

Wiley's heavier-than-air unit, impressed by the innovation he had shown in letting them off their leash, continued hooking on and dropping away to the point they did it routinely and at night.

The *Macon*'s "men on the flying trapeze" gave Wiley a trump card to play in the next naval exercise. He planned to have them carry out a dive-bombing "attack" on a carrier at night. It would have been a bold stroke but it was never to come to pass.

The *Macon* was lost in the Pacific on 12 February 1935 when its upper fin carried away off Pt. Sur, California. It rose as ballast was dropped and then descended slowly, stern down, until it splashed into the sea and disappeared in it. In contrast to the violent end of the *Akron,* its fall was almost controlled. Unlike its twin, it had life vests on board, accounting for the survival of eighty-one out of eighty-three men.

Reinforcements to Ring 17.5, which held the fins, had been decided upon after the failure-in-flight the year before en route to Opa-locka. BuAer decided that the ship had been adequately repaired and that the reinforcements could safely be delayed and done as its operating schedule permitted. Wiley, anxious to keep the *Macon* flying, concurred. Strengthening of the fins was carried out during maintenance periods at Sunnyvale. Work on the horizontal stabilizers and the lower was completed. Only the upper fin remained. To reinforce it, however, gas cells in the stern would have to be deflated and the airship placed out of service. Under pressure to satisfy the surface Navy's requirements, Wiley postponed work on the fourth fin, scheduling it for the next major overhaul. Events of 12 February showed the delay was fatal.

The Navy was not interested in looking for or recovering any of the wreck. Its whereabouts would remain unknown for fifty-five years, until 1990 when a girder pulled up by a fisherman provided the clue that led to its discovery. A manned Navy submersible, the *Sea Cliff,* found it. An unmanned remote-controlled submersible of the Monterey Bay Aquarium Research Institute (MBARI) explored and surveyed it.

Chris Grech, the driving force in getting the Navy and MBARI interested, operated the MBARI vehicle. His cameras showed the *Macon* lying on a flat undisturbed sea bottom, 1,450 feet down about three miles from shore. Its debris field, consisting of what remained, included corroded and broken girders, imploded fuel tanks, engines, mooring assembly, and the ship's control car with furniture and a pencil inside. The outer cover was missing. Some chinaware from the galley was recovered with a manipulator arm.

Most striking were the remains of the four Sparrowhawks that were on board the *Macon* at the time of the crash. They were found huddled together, laden with silt, and partially destroyed by salt water but with some fabric remaining on their wings.

The debris field from the *Macon*'s forward and midsections lies there today. A similar field, from the airship's stern, was not found.

Whatever remains marks not only the grave of the ZRS-5, but also the end of rigid airships in the U.S. Navy, then, now, and probably forever.

FIFTEEN

Farewell to Racing

In January 1934, his duties in Akron completed, T. G. W. Settle received change-of-duty orders back to Lakehurst to be in charge of its ground school.

This was an assignment "Tex" didn't want. He thought of himself as a naval officer first, an airshipman or balloonist next. He wanted sea duty. Admiral King understood and helped find him a command afloat. It was not a major one, but it was a command. He would be captain of the USS *Palos,* a three-hundred-ton, flat-bottomed gunboat of the Yangtze River Patrol in China.

Settle left Lakehurst and Navy Lighter-than-Air for the Far East in May 1934. Before he did, he made his final appearance in a Gordon Bennett race. It was in September 1933.

His 1932 victory with Bushnell was the third in a row for the United States. Van Orman had won it the two previous times. America now retained the trophy, the NAA being its custodian. The *Chicago Daily News* donated a replacement, enabling the competition to go on.

The 1933 race was at Chicago on 2 September from the Curtiss-Wright-Reynolds Airport at Glenview, Illinois. Seven balloons took part. Two were American. There was a Navy one, piloted by Settle. Because Bushnell had perished in the *Akron,* "Tex" had a different aide this time, Lt. (jg) Charles H. Kendall. A Goodyear balloon, crewed by Van Orman and a new aide, Frank A. Trotter, was also on the field.

Germany was represented by two balloons, Belgium, France, and Poland by one. DeMuyter was not the Belgian entry, his *Belgica* having become too old and porous for him to compete effectively. One of the German pilots, Fritz von Opel, was an experimenter with rocket-propelled railway cars, airplanes, and automobiles. Few of those fared well, and his balloon—although it had no rocket attachments—was no exception. It tore away during inflation, ripped itself open, rose a couple of hundred feet into the air, and then fell empty onto the field, narrowly missing the Navy's entry.

Settle and Kendall, in the four-year-old A-8476, were airborne at 6:46 PM that 2 September. Their bag was inflated with a mixture of gases: twenty thousand cubic feet of hydrogen from cylinders, sixty thousand cubic feet of coal gas from a pipeline. They wanted to keep south to avoid foul weather (thunder storms) to the north.

A southwest wind of fourteen knots started them across Lake Michigan. When they came to Lake Ontario, they drag-roped their way over it. To replenish their ballast, they hoisted water from it in canvas buckets.

Settle waves as he and Lt. Charles Kendall begin the 1933 Gordon Bennett in Chicago. (They ended up second.) The race would be Settle's last. *T. G. W. Settle Collection*

About mid-afternoon of the fourth, they came to the shore and New York State. Now the wind was thirty to thirty-five knots. Utica, Herkimer, the Hudson River, and the Berkshire Mountains passed below. Settle looked for winds blowing toward the southeast. He was thinking of crossing Long Island Sound. But he had jettisoned the balloon's air-filled pontoons as disposable ballast on leaving Lake Ontario. He also didn't want to be disqualified if he made a water landing. So he brought A-8476 to earth at Hotchkiss Grove, Connecticut, eleven miles from New Haven. He and Kendall landed at 10:45 PM, having flown 775 miles. They placed second in the race.

The Navy balloon had had an easy time of it. Not so for Van Orman and Trotter in *Goodyear IX*.

"Van" had chosen to drift north, hoping to find strong winds around Lake Superior that would take him toward Newfoundland. At first, he and Trotter flew effortlessly atop an inversion over Lake Michigan, riding it a hundred feet or so above the water. At one point, they went too low. Their basket splashed into the lake, soaking the bags of sand hanging on its sides and making them much heavier. Thanks to its pontoons, the balloon bounced off the waves and continued on. It repeated this act several times but Van Orman was able to keep it in the air.

Over land, after passing the Straits of Mackinac, he went to fifteen thousand feet where he found himself confronted by a thunderstorm. He couldn't escape it. The disturbance threw *Goodyear IX* around like a toy. He would later say that the balloon's turning, twisting, spinning, and up and down movements were more violent than those he had experienced in his ill-fated flight with Morton in 1928.

He was caught by the storm for six hours. Finding himself out of ballast, he had to land. The wind speed was fifty-five miles per hour! To slow down, he dragged the basket through the tops of the trees. They hit one and snapped it in two. They hit another with the same result. And another. And another. The fifth resisted them more. Its branches caught hold of the bag and ripped it open, leaving *Goodyear IX* dangling twenty-five feet above the ground.

Using rope from the basket, the two men lowered what it contained that might be useful to them and then lowered themselves. They were stranded in the Canadian wilderness, just as the Navy balloon crew—Kloor, Farrell, and Hinton—had been in 1920. Van Orman and Trotter, however, had some idea where they were. In the Timagami Forest, Ontario, with the nearest railroad twenty-five miles away.

Making their way through the underbrush and fallen trees was difficult beyond belief. They inched their way through it, sustained by eating from tins of food brought from the basket. One gave them ptomaine poisoning, weakening and discouraging them so much they could hardly stumble along.

After about a week of this, they came upon a high-tension line and alongside it a telephone wire carried by thin wooden poles. If they cut the phone line, someone would come to find the cause of the outage. With a small axe they had, they cut down the pole and severed the wire.

Nearby they found a linesman's shack and took shelter in it. That was after they shot the lock off the door. (Balloonists customarily carried firearms if landing in remote areas was a possibility.) Inside were some blankets, a bed, some beans, and a stove. They settled in to wait for someone to show up. When one did, he turned out to be an employee of the power company that generated and transmitted electricity from Sudbury. The ordeal was over. Altogether, they were in the wilds fourteen days. Despite what they had been through, Van Orman and Trotter were third in the race with a run of five hundred miles.

You can imagine the consternation of the Gordon Bennett officials over the disappearance of Van Orman and Trotter. They sent out aerial searches. It was suggested that the Navy send the *Macon* from Lakehurst to join in.

Another balloon was also missing. It was the Polish entry, Hynek and Burzynski. There was concern that either balloon might have landed in the sea. The Polish balloonists, it turned out, had come down ninety miles north of Quebec City. It took them twelve days to reach a populated area.

Hynek and Burzynski won the 1933 Gordon Bennett with a flight of 840 miles, the first such win by their country.

Thomas G. W. Settle and Ward T. Van Orman, America's best balloonists, would never race again. Settle was overseas on the China Station and unavailable. Van Orman, a recent widower, had three young children to raise.

The Poles, meanwhile, were on a roll. Hynek won the 1934 Gordon Bennett at Warsaw with a flight of 808 miles. He repeated his performance in 1935, again from Warsaw, making good 1,025 miles.

Balloon racing, however, was losing its appeal. It no longer drew the large crowds at takeoff. Other racing competitions—by automobiles, for example—were replacing it in the public eye. At auto races, people could see the finish as well as the start—not the case with a balloon contest.

The last James Gordon Bennett International Balloon Race the Navy participated in was the one in 1935 from Warsaw. It had one balloon entered, Lt. Raymond F. Tyler, pilot, Lt. Howard T. Orville, aide. Tyler, a former enlisted man whose lighter-than-air service dated to World War I, was one of the most experienced of the Navy's airship and balloon pilots. Orville was a meteorologist. They made a good team. But they didn't win. In fact, they finished last!

The final National Balloon Race was in 1936 from Denver. It had two Navy balloons flying in it. Lt. Cdr. Francis H. Gilmer piloted one, Lieutenant Tyler, the other. Gilmer took third place (sixty-four miles) and Tyler fourth (fifty-five miles), unimpressive mileage considering what "Tex" Settle used to rack up. Because American interest in balloon racing had declined, the yearly "nationals" were discontinued. In Europe, Gordon Bennett races were held in 1936, 1937, and 1938 resulting in two victories for Belgium and one for Poland.

In their enthusiasm for ballooning, the Poles turned to it for science in addition to sport flying. They planned a stratosphere ascent for 1939. Anton Janusz was to be the pilot.

The Poles purchased helium from the United States and shipped it to Poland in cylinders. (Following the fire that destroyed the zeppelin *Hindenburg*, Congress had enacted legislation permitting its export for scientific purposes.) The fireproof gas arrived in Poland in the spring of 1939. When Hitler invaded their country, the Poles opened the cylinders and released the gas into the air.

With the war, international balloon racing ceased altogether.

SIXTEEN

1935–1941

THE LOSS OF THE *MACON* LEFT THE NAVY WITHOUT A RIGID AIRSHIP to fly.

There remained only one such aircraft operating anywhere in the world: the *Graf Zeppelin*. Aerial globetrotter that it was, busily establishing a regular service between Germany and South America, it rarely came to Lakehurst—only five times in all.

In 1936, a new zeppelin named *Hindenburg* began landing there. Hugo Eckener, airship captain *sans pareil,* had been granted by the Navy Department a license to use the naval air station at Lakehurst as a terminal for ten transatlantic demonstration flights that summer.

The *Graf* had proven too small to be viable commercially. A larger zeppelin was needed. The *Hindenburg* was it. The 803-foot-long, 135-foot-diameter airship was the largest man-made object that had ever flown. Four Daimler-Benz diesels could push it through the air at up to eighty-two miles per hour. It had accommodations on board for fifty passengers, compared with the older ship's twenty. Because an American law prohibited the export of helium, a U.S. monopoly, the ship had had to be inflated with 7,060,000 cubic feet of flammable hydrogen.

The *Hindenburg,* Luftschiffbau-Zeppelin builder's number LZ-129, arrived at Lakehurst for the first time on 9 May 1936. Despite the enormous red and black Nazi swastikas painted on its upper and lower fins (Dr. Eckener, an outspoken anti-Nazi, despised these symbols of the Hitler government, but his airships had to carry them by order of the government), it received an enthusiastic welcome over New York.

The public took to the *Hindenburg* in a big way. The fare was $400 one way for the two-and-a-half-day crossing between the zeppelin terminal at Frankfurt and the air station at Lakehurst. Round trip was $720. Passengers liked its extraordinary quiet, absence of vibration, and lack of a feeling of motion. (Nobody got airsick!) They liked its food and drink and the smoking room (pressurized to keep stray hydrogen out). They spent hours gazing out the large, outward-slanting, draft-proof windows that lined an inside "promenade deck." The windows were kept open in flight. There was even a shower bath on the lower of the dirigible's two passenger decks. Twenty-five cabins provided sleeping accommodations. All passenger spaces were within the hull.

Eckener, ever grateful to the Navy for making Lakehurst's hangar, masts, personnel, and services available, invited its officers to fly on the ship's North and South Atlantic passages. A dozen did. One was Francis W. Reichelderfer, Lakehurst's executive officer. He heard the opening gunshots of the Spanish Civil War as the *Hindenburg* flew above Spanish Morocco in July.

Thus, during 1936, Navy Lighter-than-Air continued active rigid airship operations, although not their own. Officers flew in the *Hindenburg;* enlisted personnel ground handled it.

The success of the 1936 season led to an increased schedule of eighteen round trips the following year and an increase in passenger capacity from fifty to seventy.

After a flight in March to Rio, where the Brazilian government had provided a hangar, mast, and passenger terminal for the zeppelin service, the *Hindenburg* arrived over Lakehurst shortly after seven o'clock on the evening of 6 May 1937. It had been marking time off the New Jersey coast waiting for a cold front with thunderstorms to leave the area.

Its captain was Max Pruss. Dr. Eckener was not on board. Pruss waited for word from the air station's commanding officer, now-Cdr. Charles E. Rosendahl, that it was safe to come in and land. When NEL (Lakehurst's call letters) radioed, "Recommend landing now," he did.

The great ship emerged from the rain clouds south of the base, made a swing around toward the west and then the north, finally heading southeast to face the mooring mast. A ground crew of Navy men and civilians was waiting to receive the airship. The zeppelin came to a stop about two hundred feet in the air. Its landing ropes were dropped, the steel mooring cable at its bow began to be let out.

As the *Hindenburg* sat nearly motionless in the sky, a small flame suddenly appeared at its top, just forward of the upper fin and over a vent shaft between two gas cells.

Hindenburg at Lakehurst in 1936. *U.S. Navy Photo*

Seconds later, the after portion of the airship burst into flames. Quickly the fire burned its way forward through the hull until it emerged through the bow. The zeppelin fell to the ground while passengers and crew jumped from it and ran for safety. In less than a minute, the creation that had been the *Hindenburg* was destroyed. There were sixty-two survivors out of the ninety-seven persons on board. The thirty-six dead included a civilian member of the ground crew.

That so many on board survived can be credited to the role Navy airshipmen played in saving them. Typical were the actions of Frederick J. Tobin, a chief petty officer who had served in Navy Lighter-than-Air since *Shenandoah* days. He was the personification of a leading chief who always knew what to do. The *Hindenburg* fire was no exception.

Tobin—barrel-chested and called "Bull" because of his voice, which, according to Lakehurst legend, could be heard from one end of the eight-hundred-foot-long hangar to the other—was in charge of part of the ground party. It was under the *Hindenburg* when it caught fire. His crew broke and ran. Tobin brought them back.

Above the noise of the breaking and burning airship and the screams of those inside and outside it, Tobin's voice could be heard: "Navy men! Navy men! Stand fast! There are people in there and we're going to get them out!"

The result was near-miraculous. He got his crew together and led them into the blazing wreckage. Many who had been trapped inside were led out or carried to safety by Tobin and his crew. Other chiefs did as Fred Tobin did but without such well-remembered words.

The exact cause of the disaster was never identified and is, even now, hotly debated. The dirigible was attempting to land between thunderstorms in a heavily electrified atmosphere. A spark was generated under these conditions that ignited the hydrogen.

It had valved hydrogen just before landing. Had a valve stuck open? Had the ship, as it approached the mast, moved so slowly that its venting system was unable to expel all the valved-off gas? During its turn to meet the ground crew, did a bracing wire break and slash open a gas cell releasing hydrogen into the hull? Did an observed fluttering of the outer cover near where the flame first appeared mean that there was hydrogen loose inside that was rising and trying to escape?

As for the igniting spark, it resulted from the electrical grounding of the *Hindenburg* after it dropped its landing ropes. At first they had been dry. But as the ship paused over the mast in a light drizzle, the lines became wet and electrically conductive. Those lines were secured to the ship's metal frame and grounded it. The outer cover, however, separate from the frame, was not grounded, making for a difference in electrical charge between structure and cover. This difference generated the fatal spark.

A popular theory was sabotage. There was even the allegation that the German and American investigating committees knew that the zeppelin had been purposely destroyed and concealed the fact. South Trimble Jr., chairman of the U.S. investigators, wrote many years later that no evidence of sabotage was found. He added that they had carefully looked for evidence of sabotage in view of the feeling by Rosendahl, and some of the *Hindenburg*'s crew, that sabotage had to have been responsible.

The wreckage was transported to Perth Amboy, New Jersey, to be melted down. Significant parts were returned to Germany.

Reacting to the horror of the *Hindenburg* fire, Congress enacted legislation to permit export of helium. None was ever delivered to the Zeppelin operators. Secretary of Interior Harold Ickes blocked the sale. Rosendahl and other airshipmen raised such a furor that the secretary quipped he would in the future spell helium with two L's.

Ickes's ban held. The newly completed LZ-130, sister of the *Hindenburg*, flew in Germany in September 1938 with hydrogen. Denied helium, it would never carry passengers. After thirty flights, the LZ-130 and the retired *Graf Zeppelin* were scrapped by order of Hermann Göring, Nazi air chief, in May 1940. They were the world's last rigid airships.

Lakehurst's complement returned to its pre-*Hindenburg* routines with the K-1, ZMC-2, J-4, and a recently acquired (1935) 183,000-cubic-foot nonrigid, the G-1. Prior to becoming a Navy airship, the new acquisition had been the *Defender*, flagship of the Goodyear advertising blimp fleet. It had been christened such by Amelia Earhart. The *Los Angeles* lay decommissioned and out of service in the hangar.

After twenty-nine years (1908 to 1937) of operating more than fifty airships, the U.S. Army, in June 1937, ended its airship service. Airships and airship material were transferred to the Navy. Included were the TC-13 and TC-14, the most modern and largest nonrigids in existence.

TC-13 and -14 were sisters and nominally 350,000 cubic footers. Their cars had wheels on which they rolled to develop aerodynamic lift during takeoff. To

The Navy inherited the Army's TC-13 (shown here) and TC-14 in 1937. *U.S. Navy Photo*

keep the tail from striking the ground when the nose went up, they mounted two vertical lower stabilizers in an upside-down V-configuration. They were the pioneers of the "heavy" takeoffs that would ultimately characterize the Navy's own blimps.

They were delivered, disassembled, to the Navy in 1938. The TC-13 was in need of a replacement envelope. The TC-14, after being reerected, was ready for flight that year, thus adding another airship to the Navy's inventory.

The most noteworthy Army airship was the *Roma,* a giant semi-rigid (blimp with a keel) of 1.2 million cubic feet purchased from Italy. Prevented by inter-service agreement from acquiring an airship of rigid design, the Army had sought to achieve the size and performance of one using the semi-rigid approach. On 21 February 1921, the *Roma* buckled in flight, nosed into the ground, and struck power lines at Norfolk, Virginia. Of the forty-five Army men on board, thirty-four lost their lives. It was the deadliest American aviation accident until the USS *Akron* (eighty-one lives) in 1933.

In April 1938, the air station took delivery of a new 123,000-cubic-foot ship, a copy of the Goodyear advertising blimp. It was procured by a Navy contract with Goodyear-Zeppelin dated 11 August 1937. The same contract called for Goodyear to build a 246-foot-long, 404,000-cubic-foot patrol airship. The latter, an entirely new design, arrived at Lakehurst in mid-December 1938. The small ship, designated the L-1, was to be a trainer. The large one, the K-2, was to be a prototype patrol craft. During World War II, 22 L-ships and 134 K-ships would see naval service.

About this time, Charles E. Rosendahl, senior survivor of the *Shenandoah,* commanding officer of the *Los Angeles* and *Akron,* and, most recently, commander of Lakehurst, was detached from the air station and assigned sea duty. His next assignment ashore would be to plan the airship buildup for the war that was coming.

Between the years 1935 and 1938, Navy lighter-than-air flight training was at its nadir. There were no rigids to train for, only nonrigids. The number of trainees was small. Lakehurst's Class X (1936–37) had only two officers. Class XI (1937–38) had six.

Free ballooning continued at Lakehurst. The German zeppelin operators didn't require balloon training for their officers, but the U.S. Navy did. Its balloon training syllabus consisted of a minimum of seven free balloon flights, one of them solo and one at night.

Activity at Lakehurst in 1939 was much the same as the year before, except there was a new training class (again six officers). A replacement envelope was

Charles E. Rosendahl was the Navy's and the nation's best-known dirigible expert in the 1930s when this picture was taken. *U.S. Navy Photo*

received from Goodyear so the TC-13 could fly. And the dismantlement and scrapping of the *Los Angeles* began.

As the war clouds gathered, the Navy ordered two more L-ships and four more Ks. At the Navy Department, Assistant Secretary (later Secretary) of the Navy Charles Edison (son of the inventor) had Rosendahl, now a captain, assigned to his office when he returned from sea duty. Edison, recognizing Rosendahl's expertise and experience, wanted him to be his special assistant for airship matters.

At Lakehurst, preparations were under way for a war with Germany. Its commanding officer, Capt. George H. Mills, kept the two TC airships and the K-2 flying to develop a suitable antisubmarine doctrine for airships (there was none at the time). Practice bombs were dropped. If there were no bomb racks available, the ordnance was pushed out the door by foot.

The first three of the four K-ships on order were delivered in September, October, and November 1941.

The stage was being set for war.

SEVENTEEN

WORLD WAR II

In June 1940, Congress authorized forty-eight blimps for the Navy. It would later increase that number to two hundred.

Out of this emerged an assembly-line-produced, standard patrol airship, the "K," a derivative of the K-2 of earlier years.

One hundred thirty-four such airships would be built. Made by Goodyear, now known as Goodyear Aircraft (the Akron company dissolved its affiliation with Germany's Zeppelin Company).

The ZNP-K (Navy designation for Lighter-than-Air, Nonrigid, Patrol, K-class) was nominally 425,000 cubic feet in helium volume and 250 feet long. It had a three-ply rubberized cotton envelope with air ballonets inside to control internal gas pressure. Powered by twin Pratt and Whitney R-1340-AN2, 425-horsepower engines, it had a maximum speed of 67.5 knots and a cruising speed of 50 knots. Range at cruising speed was 1,900 nautical miles.

"K's" were deployed, six or twelve to a squadron, at the new lighter-than-air stations built for them at South Weymouth, Massachusetts; Weeksville (Elizabeth City), North Carolina; Glynco (Brunswick), Georgia; Richmond (Miami), Florida; Houma, Louisiana; Hitchcock, Texas; Santa Ana, California; and Tillamook, Oregon. Each of these stations would have one, two, or three hangars that were a thousand feet long. They were the product of Rosendahl's planning while he was a member of Secretary Edison's office.

In addition, there were the existing bases at Lakehurst and Sunnyvale. Each possessed a rigid airship hangar. Lakehurst had been in continuous operation by the Navy since the 1920s. Sunnyvale, originally the home for the *Macon,* had been transferred to the Army in 1935. It would be returned to the Navy, which

L-ships (training and patrol), G-ships (training and utility), K-ships (patrol), and M-ships (patrol) saw naval service in World War II. *U.S. Navy Photo*

would rename it the Naval Air Station, Moffett Field, for its BuAer chief who died in the *Akron*. Lakehurst and Moffett would each have two of the new thousand-foot hangars added for blimp squadrons that would be operating there. Each would also be equipped to serve as a major overhaul and assembly site for its coast.

Airships "went to war" on the East Coast on 2 January 1942 with the commissioning of Airship Patrol Squadron 12 at Lakehurst. It had four K-ships (K-3 through K-6).

Operations began officially on the West Coast from Moffett Field on 2 February 1942 with the TC-14. It had been shipped there by rail from Lakehurst to be assembled and inflated. The Pacific's first operational blimp squadron, Airship Patrol Squadron 32, commissioned at Moffett on 31 January, operated at first using the ex-Army TC-13 and -14 and small L-type advertising blimps requisitioned from Goodyear.

To its crew of four officers and six enlisted men, the K-ship was a reliable, pilot-friendly, and forgiving aircraft. Its ability to fly long, low, and slow, and to keep pace with vessels it was escorting made it well suited to antisubmarine missions.

The best known airship picture of the war: a K-ship shepherding a convoy. A good-sized enlargement of this photo hung over the bar in the Lakehurst Officers Club, giving rise to frequent and sometimes furious debates whether the blimp was headed toward or away from the camera. The question was never resolved. *U.S. Navy Photo*

At the very first, the "K" was limited by a lack of sensors, its prime one being the "Mk I eyeball." When a submarine contact was made, it was usually visual: the feather of a periscope moving through the water, the swirl left behind by a crash-diving U-boat, or the hull or conning tower of the submarine itself.

But not for long. Under development at Lakehurst was the Magnetic Anomaly Detector (MAD), which could detect a submarine that was underwater. Consisting of a magnetometer with a vertical range of four hundred feet, it sensed the distortion of the earth's magnetic field by the presence of a large metallic body, such as a U-boat hull. Owing to its limited range, it was not useful for a general search. But it could be invaluable for confirming and developing a visual sighting. MAD could not distinguish, however, between a submarine, a wreck, or a geological formation, so it was prone to false alarms.

In July 1942, another sensor was added. Radar! K-ships began carrying a plan-position-indicator type, which displayed maplike returning echoes on the scope. With radar, blimps began escorting ships day and night.

During those first few months, airships made and bombed numerous contacts but without proof of success. It would not be until April 1942 that a U-boat was destroyed in American coastal waters by the destroyer *Roper*.

As its sensing capabilities increased, so did the K-ship's armament. There was a .50-caliber machine gun in a turret over the pilot's head and a lesser one in the after end of the car. The K-ships carried depth bombs (four maximum). Contact bombs—shipboard Hedgehogs with their propellant removed—would be added to drop directly upon sources of MAD contacts. Later on, so would be the ultimate U-boat killer, "Fido," the Mk 24 mine, an acoustic torpedo. Because the Mk 24 homed on the cavitation sounds from the screws of a submerged submarine, blimps carried sonobuoys—small, radio-equipped, listening buoys—to be dropped when determining whether a U-boat was present or not.

As their numbers increased, the "K's" began filling up the hangars that had been built for them. By 1943, their numbers had increased to the point that the time had come to establish administrative "type commanders" for the airships in the Atlantic and Pacific. Day-to-day operational control of the squadrons would continue to be exercised by the Sea Frontiers, Fleet Air Wings, and others.

On 15 July 1943, a Fleet Airships, Atlantic, command was created, headquartered at Lakehurst, with eleven patrol squadrons, henceforth to be called "blimprons." Servicing and maintenance would now be performed by blimp headquarters squadrons ("blimphedrons"), like the blimp squadrons part of Fleet Airships, Atlantic, and its four administrative wings.

The commander of Fleet Airships, Atlantic, would be George H. Mills, Lakehurst's former commanding officer, who had organized the airship antisubmarine effort in the Atlantic from the very beginning. He would continue to do so for the duration of the war with Germany. Mills was promoted to the rank of commodore.

On the West Coast, at Moffett Field, a similar command, Fleet Airships, Pacific, was simultaneously created. It had one wing comprising three blimprons and a blimphedron. It was headed by an officer of captain rank. The first was "Tex" Settle. (Settle wanted duty in command of a combatant ship in the Pacific. [He had found his "sea legs" as a destroyer officer early in his career.] He was later able to break away from Lighter-than-Air to command the heavy cruiser *Portland,* in which he distinguished himself in combat with the Japanese and earned the Navy Cross. In postwar years, as a vice admiral, he would be Commander, Amphibious Forces, Pacific Fleet.)

Another major change in Navy Lighter-than-Air took place on 15 May 1943. The Naval Airship Training and Experimental Command was formed, based at Lakehurst, and under a rear admiral, Charles E. Rosendahl. He had returned from sea duty after leaving the Navy Department in early 1942 to command the *Minneapolis.* Like Settle, he returned from the Pacific with the Navy Cross, having saved his shot-up cruiser.

Under Rosendahl's control, the training of pilots (naval aviation cadets, recalled Naval Reservists, and others transferred into Lighter-than-Air) was divided. Primary flight training would be given at Moffett, advanced at Lakehurst. Enlisted ratings were to be trained at Lakehurst, Moffett, and other airship commands and by service schools in the case of technical specialties.

Altogether fifteen hundred pilots would be trained and three thousand enlisted aircrewmen qualified. L-ships (123,000 cubic feet) and G-ships (196,700 cubic feet) were used for primary flight training. For advanced training, K-ships.

Free balloons were flown at Moffett Field and Lakehurst. All were inflated with hydrogen to save helium. In deference to hydrogen's flammability, there was said to be a sign at Lakehurst: "No smoking! The ashes you drop may be your own!"

Despite many balloon flights, only one fatal balloon accident happened during the war. A Lakehurst balloon, making a normal landing, suddenly burst into flames. The burning bag fell over the basket, killing the three men on board. Static electricity, from pulling the rip cord, had ignited the gas.

The student officers and aviation cadets trained in balloons during the war years, being young and uninhibited, found ballooning ready-made for high jinks. They enjoyed games like landing after a snowfall, putting one of their number over the side, and leaving footprints for locals to wonder at—footprints that

Moffett Field training airships (L-type) pass in review. *U.S. Navy Photo*

began from nowhere. Or like dropping sand on love makers in isolated places and clearing out a lovers lane at night by trailing the drag rope across the roofs of the parked cars. Enthusiastic young men in balloons found it great fun bouncing off the tops of clouds, crashing through the tops of trees, and drifting cross-country in an aircraft that made no sound. They liked to watch the drag rope to see what it would attach itself to, wrap itself around, and pick up. It was great fun drifting along, noiselessly, feeling little or no wind in their faces, regardless of ground speed. Balloons were part of the atmosphere and moved along with, not through, it.

As the Battle of the Atlantic progressed, Germany's "gray wolves" moved into the Caribbean and along the coastline of South America. The blimp squadrons of the Atlantic Fleet followed them. A squadron was organized and dispatched to Trinidad, with detachments in what was then British and Dutch Guiana. Another squadron went to Jamaica, with detachments in Panama. It covered the western Caribbean and the approaches to the Canal. Farther south, two squadrons operated along the coast of Brazil as far south as Rio de Janeiro.

Launching a free balloon: First, lay out the envelope, net, and sand bags. Insert and secure the maneuvering valve at the apex. *U.S. Navy Photo*

With the net spread over it, inflate the envelope through a fabric sleeve called an appendix. Restrain the balloon by attaching the sand bags to the net and lowering them as inflation proceeds. *U.S. Navy Photo*

The gassing completed and the sand bags hanging from the bottom of the net, transfer the bags to the basket, which is being attached by foot ropes and a load ring to the net. *U.S. Navy Photo*

All this was achieved essentially without benefit of hangars. The only ones in Latin America were a small ex-Army one in Trinidad and the former zeppelin shed at Santa Cruz, outside Rio. Navy airshipmen had learned by doing that K-ships could be operated routinely in unfriendly tropical climes with little more than a mast, a mooring circle, spare parts, and helium. Repairs, except for the most major, could be made outdoors while the airship was moored out on the field. The ten or so detachments from which Squadrons 41 and 42 flew in Brazil gave proof of this.

Various areas of the Atlantic were "hot spots" insofar as the presence of U-boats was concerned. The Straits of Florida was one. There an unusual running gunfight took place between a Navy blimp and a surfaced German submarine.

It was 18 July 1943, just before midnight. The K-74, Lt. Nelson G. Grills, pilot, investigated a radar blip that turned out to be the U-134. Sighted visually as well as by radar, it was headed right for a tanker and freighter that were in the vicinity. To protect them, Grills attacked the German. (By doing so, he violated doctrine, which called for him to stay upwind of the target and radio for air and

Walk the balloon to its release position. Give attention to the appendix dangling below the envelope. It was tied closed when inflation was completed. Before leaving the ground, it must be untied because it serves as the envelope's pressure relief valve, venting its gas as it expands. Finally, "weigh off" the balloon, making certain it is light enough to rise, and order the ground crew to let go. *U.S. Navy Photo*

surface support.) At forty-seven knots ground speed and an altitude of five hundred feet, Grills charged the U-boat, with his .50-caliber machine gun blazing. The U-134 changed course to align its stern with the oncoming airship and replied with its own 20-mm fire and perhaps (not certain) a round or two from its deck gun.

The K-74 passed directly over the submarine. According to Grills: "I could have spit on its deck." But his bombs hung up and didn't release.

The 20-mm fire had bracketed the airship's car, setting the starboard engine aflame. A round or two from the enemy's deck gun may have holed the bag. Riddled with holes, the envelope lost its pressure, shape, and lift as it fell tail-first into the sea. One man died in the water. The other nine were rescued the following day.

The U-134 submerged. It radioed U-Boat Command that it had shot down a naval airship but had suffered damage to its quick-dive tanks and was returning home. On its way back, British planes caught it off Spain and sank it.

In 1944, Lighter-than-Air's most ambitious overseas deployment took place. A squadron was moved to the naval air station at Port Lyautey in French Morocco.

Although MAD had originally been developed for blimp use, other Navy aircraft carried it, too, notably the PBY Catalina flying boats. Those that did were called MADCATS. A squadron of these planes was flying a day and night magnetic barrier patrol across the Strait of Gibraltar to detect German submarines entering the Mediterranean (once inside it, they never left).

MAD, with its four-hundred-foot vertical range, required that the Catalinas fly close to the surface of the water. This was difficult by day, almost impossible and exceptionally dangerous at night.

Rosendahl, who handled the Washington side of airship matters, proposed blimps be substituted for the Catalinas in nighttime operations. Admiral King, the Navy's supreme boss, approved.

Six K-ships were flown across the Atlantic with refueling stops at Argentia and the Azores in May and June. Two more would follow the next year in April and May, via a Bermuda-and-Azores route. After hostilities ended, the airships of Blimp Squadron 14, "the Africa Squadron," began mine spotting in the Mediterranean.

Back home, on 5 May 1945, Lakehurst airships joined with Navy surface units in attacking the U-853 off Block Island. It would be the last German submarine sunk in American waters during the war.

The speed with which a blimp squadron had been deployed to Africa gave rise to another overseas assignment, this one to England. Blimp Squadron 42 was ordered to Chivenor in Devon to operate with Britain's Coastal Command. Airships had shown themselves adept at detecting the snorkels (breathing tubes) of submarines underwater. They were wanted for Britain's southwest approaches. Four "K's" were readied for this overseas assignment and two were on their way across the Atlantic when Germany surrendered. The deployment was cancelled. The two ships returned to the States.

Between 2 January 1942 and 15 May 1945, airships of the Atlantic Fleet had made 37,554 flights, been airborne 378,237 hours, and safely escorted seventy thousand surface vessels. They had constituted eleven patrol and escort squadrons plus one utility squadron.

In the Pacific, the corresponding figures (31 January 1942 to 1 September 1945) were 20,156 flights, 167,291 hours, and eleven thousand ships. With few, usually no, Japanese submarines in their operating area, blimps of Fleet Airships, Pacific, were denied a chance to get at the enemy. This did not prevent them from carrying out a number of spectacular air-sea rescues, as well as a rescue in the California desert. A landing by a West Coast K-ship on the *Altamaha,* the first carrier landing ever by a U.S. Navy blimp, would pave the way for an at-sea operation that became routine in postwar years.

It was a Moffett Field airship that would write a unique chapter in naval aviation history. It returned from patrol with no one on board!

The L-8, a former Goodyear advertising blimp, took off from Treasure Island in the San Francisco Bay area on 16 August 1942. Two officer-pilots were on board. Five hours later, it came to earth in the streets of Daly City, California, having drifted ashore with its crew, Lt. Ernest D. Cody and Ens. Charles E. Adams, missing. What happened to them remains a mystery. Probably one of them went out onto an engine outrigger to fix a problem. The other went to his aid and they both fell into the bay. They were wearing life jackets but neither body was ever found. Cody and Adams were among the seventy-five or eighty Navy airshipmen who lost their lives in airborne and ground-based accidents during the war.

Four types of nonrigids were operated by Atlantic and Pacific squadrons and by the Naval Airship Training and Experimental Command. The K-type was the workhorse of the antisubmarine squadrons. It was also the airship used for advanced training. (The K-2, first in the class, was set aside for research and experimental work.) The "M," half again the volume of the "K," was also a patrol airship. Twenty were ordered but only four built, the K-ship having proved itself capable of doing what had been expected of the "M." The M-1, like the K-2, was reserved for experimental projects.

With no one on board, L-8 drifts toward a landing in Daly City, California. The disappearance of the two pilots on board has never been explained. *U.S. Navy Photo*

For primary training, L-ships were used, built for the Navy or pressed into service from Goodyear. There were twenty-two of them.

There were also seven G-ships, slightly larger than the G-1, acquired by the Navy in 1935. Originally built as trainers, they formed the core of Fleet Airships, Atlantic's, "utron" (utility squadron), which performed tasks including the chasing of practice torpedoes for U.S. submarines. Each of these airships was built by Goodyear Aircraft.

No one knows the number but survivors by the hundreds were rescued thanks to Navy airships. There were fifty occasions in 1943 when victims of torpedoed ships and of downed aircraft were found. As many as fifty victims of a torpedoing were sighted at one time. In the case of airplanes forced down in inaccessible areas, K-ships set down on their landing wheel and, while their crews ground handled the ship, retrieved the downed airmen.

No one doubted the value of blimps for these utility and rescue roles. But how about their performance against the enemy?

At the war's start, the airship was widely touted as a superb one-on-one hunter and killer of U-boats. Experience proved otherwise. It could be that, but not one on one. Its antisubmarine value depended on its operating as part of a team, whether it was paired with destroyers, destroyer escorts, frigates, or airplanes. The cited incident of helping destroy the U-853 was both an example and typical.

The blimp's major contribution was "keeping the enemy under," preventing him from attacking. The disappearing radar blips picked up by airships were surfaced submarines that were diving to escape detection.

In a letter he wrote to me in March 1964, Grand Admiral Karl Doenitz pointed out the importance of making a U-boat captain submerge: "Each air patrol was disturbing to the operation of U-boats because they were forced underwater, making it no longer possible for them to reach their attack positions. As a consequence, the American 'blimps' were very disturbing to German U-boat activity. Naturally airplanes, compared with 'blimps,' had the advantage of higher speed. Thus they represented for U-boats a greater danger than 'blimps.'"

Thus wrote World War II's commander of U-boats.

EIGHTEEN

Postwar

World War II had ended when lighter-than-air suffered a major catastrophe. The naval air station in Richmond, Florida, was destroyed—including all three hangars—by a hurricane on 14 September 1945.

Airplanes, naval and civilian, had been evacuated to Richmond (southwest of Miami) to be housed in the safe haven of its thousand-foot wooden hangars. Twenty-five blimps, some in storage, were also inside. The 140-knot winds beat the structures to pieces. They caught fire. The 25 blimps, 31 non-Navy U.S. government airplanes, 125 privately owned ones, and 212 Navy heavier-than-air craft were consumed. Thirty-eight Navy men were injured. The civilian station fire chief was killed.

Loss of Richmond was an unexpected blow to Navy airshipmen. More expected, but a blow nonetheless, was the postwar disestablishment of Fleet Airships, Atlantic and Pacific, and their wings, blimp squadrons, and blimp headquarters squadrons. What remained was not much.

Numerous airships and free balloons were declared excess to the Navy's needs and offered for sale. The airships were bought by Howard Hughes and by the Douglas Leigh Sky Advertising Company. Hughes employed his as a flying billboard to advertise Jane Russell's movie *The Outlaw*. Leigh, the advertising sign king of Times Square, famous for his smoke-ring-blowing Camel cigarettes spectacular, used his "K's" and "L's" to call attention to a variety of products, including Ford cars, R and H Beer, Flamingo orange juice, and Wonder Bread. (I have a special affection for the Flamingo ship, with the illuminated flapping wings on its sides. In it I proposed to my future wife, Joanne, while over New York City.)

As far as is known, none of the surplus free balloons was ever flown. They were scrapped and their envelopes cut up for use as covers and tarpaulins.

Six thirty-five-thousand-cubic-foot balloons, with envelope, valve, bag, net, and basket were put on sale at Lakehurst in 1946. At the time, I was active in the NAA, trying to revive the prewar Gordon Bennett races, and saw them as a means to jump-start the resumption of the competition in the United States. I purchased all six. A Lakehurst shipmate found a place to store them in the station's rigid airship hangar, pending locating a permanent home.

I had begun to have serious doubts about the practicality of my purchase, given the cost of transporting, inflating, and insuring the balloons, when I learned that the air station was about to have a "field day" to clean out the hangar. I returned the balloons to "war surplus" where they met a nonflying fate. What had been a once-in-a-lifetime opportunity to acquire balloons for balloon racing had been lost.

The K-ship emerged from the war with a reputation that made the Navy want to continue its operation. Modernized and improved versions—the ZP2K, 3K, and 4K—made their appearance.

All were antisubmarine airships. They operated in close cooperation with carriers. Their crews became experienced hands at making carrier landings and at refueling from carriers by pumping up gasoline while they hovered above. Also for refueling, a carrier could deploy bladders filled with "avgas," aviation gasoline, that would float until retrieved and hoisted on board by the blimp. A method, too, was developed to exchange crewmen by a kind of basket-like car.

Avionics having become "the thing" for detecting a submarine, the newer "K's" gave ever more room to such equipment—which included towed sonar. Envelope volumes were increased to 527,000 cubic feet to accommodate the greater load.

The ZP5K bore little resemblance to any of its predecessors. Its car was longer. Instead of four stabilizing fins, it had three, its tail assembly arranged in an inverted Y. Eighteen of these 650,000-cubic-foot ships were built.

An entirely new generation of Navy blimp was born in June 1952 with the delivery to Lakehurst of the ZPN-1. It had many distinctive features, including a double-deck car and engines (two) mounted inboard. Seakeeping and habitability played an important role in its design. Its cockpit borrowed heavily from heavier-than-air practices.

Twelve ZPG-2s, 975,000 cubic feet, followed. They reflected but did not exactly copy the ZPN-1. They would be the most versatile airships of the postwar period. The ZPG-2s enjoyed many achievements.

The Navy built and operated a half-dozen nonrigid airship types in the postwar years. The best performer was the ZPG-2, shown here above the ZPG-2W, its early warning version. *U.S. Navy Photo. Courtesy Navy Lakehurst Historical Society*

For example, in May 1954, with Cdr. M. Henry Eppes as pilot, one flew 200.2 hours without refueling. Not to be outdone, ZPG-2 serial 141561, commanded by Cdr. Jack R. Hunt, took off in March 1957 from South Weymouth, Massachusetts. It flew to Portugal, then to the Canary and Cape Verde islands, and on to Morocco. When it landed at Key West, it had clocked 264.2 hours and flown 9,448 miles without refueling.

Another ZPG-2, making use of refueling stops, penetrated the Canadian Arctic in 1958 to make an air drop to scientists on Ice Island T-3. The airship was commanded by Cdr. Harold B. Van Gorder.

To convince skeptics that airships were not fair-weather aircraft, ZPG-2s undertook long-endurance patrols off New Jersey in the worst winter weather they could find. In a February–March period, they maintained station almost 90 percent of the time.

As impressive as this was, doubts continued in the Navy about the usefulness of blimps for antisubmarine warfare. They were considered too slow. In Navy

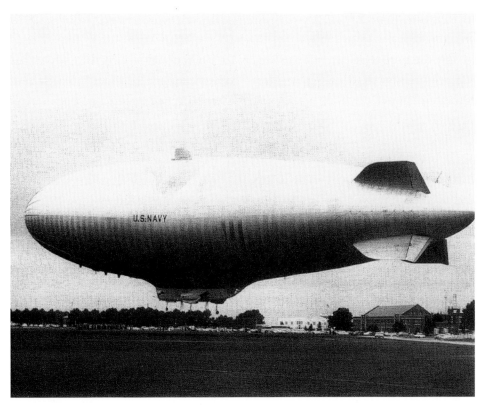

The Airborne Early Warning ZPG-3W, 403 feet long and a nominal 1.5 million cubic feet in volume, was the largest blimp ever. Four were built. *U.S. Navy. Courtesy Navy Lakehurst Historical Society*

parlance, "They can't reach a datum point fast enough." Also with the underwater speeds newly possible with nuclear submarines, an enemy could outrun an airship by making it buck headwinds. These views were the handwriting on the wall that Lighter-than-Air's days as an antisubmarine weapon were numbered.

Despite this, the airship's ability to maintain station for protracted periods was recognized, so much so that a new series, the ZPG-2W and ZPG-3W, were built. Their role would be airborne early warning, and an early warning squadron would be formed at Lakehurst to develop the mission.

The "2W" was an outgrowth of the ZPG-2 with a height finding radar atop the bag and a search radar beneath the car in a dome. Like the ZPG-2, the bag size was 975,000 cubic feet.

The "3W" was half again bigger (a monstrous 1,465,000 cubic feet). It was 403 feet long and the largest nonrigid airship type ever built. Like the "2W," it had a height finder radar on its back. But instead of a radar on the underside of the car, it mounted one—with a forty-foot-diameter rotating antenna—inside its envelope. Four ZPG-3Ws were ordered and delivered to the Navy in 1959 and 1960.

By the time these aircraft were ready for operational service, the early warning requirement had changed. They were no longer considered important. Other and better methods for detecting Soviet surprise attacks had been developed. The ZPG-3Ws were assigned other tasks.

The end was in sight for Lighter-than-Air on 6 July 1960 when ZPG-3W, serial 144242, lost its hull pressure and crashed into the Atlantic off New Jersey. It took the lives of eighteen of its twenty-one crew. Many airshipmen believed that this accident would bring an end to their program. It did. In June 1961, the airship program was ordered terminated by the Secretary of the Navy.

The Navy's last airship flight was made by a ZPG-2 at Lakehurst on 31 August 1962. A naval flying service that had begun with the DN-1 in 1917 ended after forty-five years.

It was all over.

EPILOGUE

While the Navy was winding down and ending its airship program, it was developing manned scientific ballooning.

Through the years, balloons, except for small rubber ones, had been made of cotton or silk fabric, rubberized or shellacked to make them gastight. It had taken two separate components, the envelope material and the sealant, each heavy, to make one. Balloonists were eager for something that could both carry the load and hold the gas.

Chemical engineer Jean Piccard had made small balloons out of cellophane in the 1930s but they had failed, unable to withstand cold temperatures.

Piccard was also interested in clustered balloons. Several small balloons, he believed, were superior to one large one. To prove his idea, in 1937, with a cluster of ninety-two rubber weather balloons, he ascended to several thousand feet. He called his balloon assembly *Pleiades*. The coming of war and lack of sponsor kept him from developing the idea.

The U.S. Navy would give him his opportunity.

In the mid-1940s, in Minneapolis, where he was teaching at the University of Minnesota, Piccard met a German emigrant engineer, Otto C. Winzen. Together they worked out a plan for a balloon flight into the stratosphere. It would be achieved by one hundred balloons arranged in a large cluster. A crew of two, Piccard being one, would fly it to one hundred thousand feet.

They succeeded in interesting General Mills, Inc., also of Minneapolis. Usually thought of in terms of breakfast cereals and Betty Crocker, the company had engaged in military work during the war and developed a respectable capability in mechanical engineering. General Mills was an expert in packaging and a balloon was a package of gas. It agreed to produce the balloons, provided a contract could be obtained to build them.

Winzen approached the Navy's Special Devices Center, a part of the Office of Naval Research (ONR), located at Port Washington, Long Island, New York. He met with a lieutenant commander, George W. Hoover, an imaginative and innovative aviator who headed the center's flight section. Hoover was interested in flight at high altitudes. He saw immediately that what Winzen and Piccard were proposing would give him some of the information he was seeking. Hoover arranged for an ONR contract with General Mills, named the project *Helios,* and looked for an experienced Navy balloonist to be its pilot.

Hoover selected Lt. Harris F. Smith, a Naval Reserve officer who had served in airships during the war and had a reputation for doing exceptional things with free balloons. He was the closest that the Navy had to "Tex" Settle since Settle.

Smith, who liked to practice his airmanship by writing his initials in the Lakehurst sand with his airship handling lines, had been a flight instructor there. His ballooning students had included a U.S. Coast Guard class and a Brazilian Air Force class. Both services had intended to initiate an airship arm, the Coast Guard for air-sea rescue, the Brazilians for antisubmarine operations off their coast. Neither would. The Coast Guard's plan to use blimps was sidetracked by the helicopter; the Brazilians cancelled out because the war ended.

On 20 May 1945, in balloon ZF48239, with a basketful of Brazilian trainees, Lieutenant Smith cast off from Lakehurst. When they landed, it was in the quadrangle of the Georgian Court College in Lakewood, New Jersey. A class of Navy WAVES was in training there. The arrival of the balloon ended all class work. The young women rushed out to greet their visitors. Brazilian Lt. Adolphus Meyer, who spoke fluent English, lost no time in dating the cutest WAVE. Smith would train eleven officers and thirteen enlisted men in free ballooning before Brazil's program ended. The group's commanding officer failed to graduate and departed in a huff.

Five days later—25 May 1945—Smith was airborne again with four enlisted Coast Guardsmen. The balloon's drift was north toward New York City where it astonished office workers and others by floating over Twelfth Avenue, along the Palisades, and up the Hudson River. When it came to Ossining, it overflew the yard of Sing Sing prison. Shouts of "drop us a rope" were heard from below. The flight was 3.7 hours long. It ended atop a tombstone in a graveyard at Kent, Connecticut. A local family fed the aeronauts. The chase truck from Lakehurst, after running out of gasoline and money, arrived at midnight. The balloon was loaded onto the truck, which Smith drove in the wrong direction for an hour while the others slept. When he realized his error, he turned, and they were back

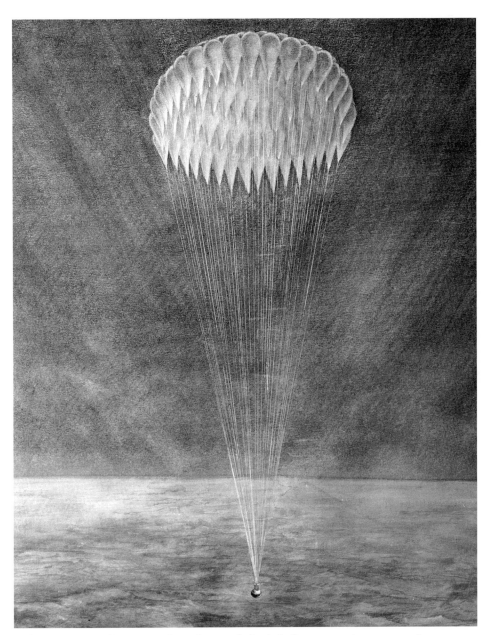

Helios, Jean Piccard's concept for a cluster of plastic balloons carrying a crew of two at one hundred thousand feet. *U.S. Navy Photo*

at Lakehurst before breakfast. For taking the balloon out of New Jersey, Smith would be placed "in hack"—restricted to quarters.

In October, on the tenth, he and fellow airshipman Lt. William J. "Beaver" Gunther made a night flight that carried them into the New York–Washington airway. The airlines were notified. Beneath its basket, the balloon carried a white

Lt. Harris F. Smith, the balloonist originally selected by the Navy to pilot *Helios*. He is shown here (right) with Lt. William J. Gunther. Their long-endurance flights together in 1945 reawakened Navy interest in ballooning. *Harris F. Smith Collection*

running light and red warning light. There was also a flashlight to shine against the bottom of the bag when an approaching airplane was heard. It would identify what was in the sky.

At first, their track was southwest to the Susquehanna River. Then the wind carried them north and back toward Lakehurst. For a while, it appeared they might land where they took off. Smith even radioed Lakehurst an estimated time of arrival of 5:05 PM.

But the wind died. The dream of all free balloonists—to return and land where they started—was not to be. Smith and Gunther landed at Whiting, in a wildlife management area, two miles from the air station. Probably never in the history of free ballooning had a flight of 25.4 hours come so close to returning to its takeoff point!

On 30 October, Harris and "Beaver" took to the air at four o'clock in the afternoon to examine the feasibility of making overnight balloon flights a routine part of the syllabus for the Officers Airship Training School. There was another reason. It was 1945. Hopes ran high that international balloon racing would resume and that the Navy would participate. Smith and Gunther were expected to provide experience useful in such competitions.

When they took off, Lakehurst Aerology predicted a landing for them the next morning in the vicinity of Newark, New Jersey. They had sixty pounds of sand ballast to get them there.

From Lakehurst, they were carried northwest over Allentown, Pennsylvania, and then Binghamton, New York. By four o'clock in the morning, they were over Montreal. Their radio couldn't raise the Montreal tower but they were still in contact with Lakehurst's.

There was lightning, thunder, and rain. Their ground speed reached eighty miles per hour! While "Beaver" was asleep in the sack, Smith, who had a sore throat, was feeling no desire to spend days wandering in the Canadian wilderness, even though they had six days of emergency rations on board.

When Gunther awoke, he and Smith conferred on what to do. "Beaver" was all for continuing as long and as far as the balloon would take them. Smith, who had checked out balloon ZF48241 and was responsible for it, was more cautious, preferring to come down when they reached the edge of the maps they were carrying. Or until they came to a suitable, preferably populated, landing area.

According to Smith:

> At 0600 I spotted the International Paper Company's pulp log pond at La Tuque, Quebec. We were about 75 miles from Quebec City.
>
> We missed making a landing in the pond and were blown up the Maurice River gorge. The wind bounced us off its east cliff. I valved us down into the river which we hit at about 40 knots.
>
> Our pontoons heeled the basket over with the top of the balloon almost touching the water. Bill and I stood on the rear edge of the basket to keep its leading edge from scooping up water. Suddenly we became aware that our life jackets were still in the bottom of the basket and that the river was full of 10-foot pulp logs. We managed to put the jackets on while holding onto the basket's ropes with one hand. The wind across the gorge gradually carried the balloon to the west bank. "Beaver" threw his spare pants, with his wallet inside, onto the bank. I ripped the balloon. The basket sank under water. Four homing pigeons drowned.
>
> We escaped the balloon as it was being swept downstream. We climbed up the west bank. "Beaver" retrieved his trousers and wallet.
>
> Then we started walking along a railroad track against the wind. We came upon two men riding a fully loaded, motorized, hand car. They refused my high-school French request for a lift.
>
> After walking an hour in our wet winter flight gear, we crossed over the Maurice on a railroad bridge. We stopped at a farmhouse where I called "Rosie" [Admiral Rosendahl] at Lakehurst, using a hand-crank phone. La

Tuque police came to collect us and took us to the local community club. Its manager put us to bed with a half-bottle of Canadian Club and instructions to drink and sleep while our clothes dried.

I slept until 10:00 PM, then went down to a Halloween party in process. "Beaver" had been in La Tuque, spending his dollars.

We boarded a sleeper train to Montreal where "Rosie" had arranged a plane ride for us on a diplomatic flight to Washington which dropped us off at Lakehurst.

The flight had covered 1,120 miles in fourteen hours. It was the longest by a Navy balloon since Settle and Bushnell won the Gordon Bennett in 1932. It was vintage Settle, vintage Van Orman, and vintage DeMuyter all the way.

This was the unflappable balloonist and lieutenant named Smith who was tapped by the Navy to pilot *Helios*. A Princeton graduate, he was accomplished in diplomacy and personal relationships. He had to be!

A flight to the top of the atmosphere and a world's altitude record was a glittering attraction. It brought engineers, physicists, biologists, medical doctors, and others out of the woodwork, all wanting a piece of the action. Lt. Cdr. George Hoover, the project's sponsor at the Special Devices Center, took to calling it Project Durante, recalling that comedian's line that "everybody wants to get inta da act."

Those who became involved included Navy "brass" who allowed Smith to be elbowed out and replaced by a Navy commander with no record of significant ballooning achievements.

Helios never flew. The one-thousandth-of-an-inch-thick polyethylene film balloons proved unreliable. The difficulties in simultaneously launching one hundred ten-thousand-cubic-foot balloons were staggering. *Helios* was cancelled. A program of unmanned, single-cell, plastic film research balloons, to be called *Skyhooks,* was substituted.

These twenty-thousand-cubic-foot polyethylene cells—they would later grow into millions of cubic feet and be made of other plastic material—were welcomed by scientific investigators as a comparatively inexpensive way to collect data at very high altitudes. Emulsions and films to show the passage through them of incoming cosmic rays were typical payloads sent up.

As the reliability of these unmanned balloons improved, the question arose: why not a human payload?

Charles B. Moore, a meteorologist who had succeeded Winzen at General Mills, put the idea to a test on 3 November 1949 when he attached a parachute to a small polyethylene cell. Making himself comfortable in its harness, using sofa cushions from home, he took off over the Minnesota countryside. He

Some of the cast of project *Helios:* Lt. Cdr. George W. Hoover, prime mover for the Navy Special Devices Center, is in the center, in uniform, standing beside Jeanette and Jean Piccard; the Piccards are next to Otto C. Winzen. Lt. Harris Smith is last on the right. *U.S. Navy Photo. Courtesy Harris F. Smith*

landed safely, his test a success. Moore had opened the pages of a new chapter in ballooning history.

Another meteorologist, Malcolm D. Ross, a lieutenant, later a commander, attached to the Office of Naval Research (Air Branch) was also interested in manned flights.

So was M. Lee Lewis, another lieutenant, from the Aerology Office of BuAer. He was based in Minneapolis as the Navy's local representative for the balloon work going on there. Gentle, quiet, and modest, he was a beloved figure in ballooning. He would retire as a lieutenant commander and join Winzen Research, Inc., a balloon manufacturing firm formed by Otto Winzen after leaving General Mills. (Lewis would be killed in a freak accident. While working indoors on a gondola suspension, he was struck by a pulley block that broke away.)

On 8 November 1956, lieutenant commanders Ross and Lewis, in a pressurized spherical gondola originally built for *Helios,* surpassed the existing record height for manned flight of 72,395 feet. It had been achieved by *Explorer II* and Army captains Albert W. Stevens and Orvil Anderson on 11 November 1935. The two Navy men reached an unofficial 76,000 feet.

In 1934, two American balloons had climbed to the stratosphere: *Explorer* and a second one flown by Jean Piccard and his wife, Jeanette. They had received

Charles B. "Charlie" Moore made the first manned flight with a polyethylene balloon on 3 November 1949. *Author's Collection*

title to the Settle-Fordney balloon and reconditioned the bag and gondola. They took off from Dearborn, Michigan, Jeanette as pilot, Jean as observer, and a turtle as mascot. The three reached 57,579 feet on 22 October. (If ever there was a royal family of ballooning, the Piccards were it. Auguste had twice ballooned into the stratosphere. Donald Piccard, Jean and Jeanette's son, flew in a former Japanese wartime bomb-carrying balloon. A relation, Bertrand Piccard, with Brian Jones, flew 29,054 miles in 1999 to circumnavigate the globe.)

Some weeks earlier—28 July 1934—the U.S. Army Air Corps/National Geographic Society balloon, *Explorer,* had risen from a natural hill-surrounded bowl, dubbed the Stratobowl, in the Black Hills of South Dakota. Maj. William E. Kepner, Capt. Albert W. Stevens, and Capt. Orvil Anderson were on board. The flight went smoothly until sixty thousand feet. Then the envelope began to tear and rip apart. The *Explorer* began to fall. Like a supersized parachute, the bag lowered the gondola toward the ground. A mile above it, as air mixed with the hydrogen, the envelope exploded. Now the gondola and those inside were in free fall. The three were barely able to escape through the hatches and parachute to safety.

Ross and Lewis, in reaching seventy-six thousand feet, also had tense moments. Theirs was the first in a series of very high flights called Strato-Lab.

Charlie Moore, right, tending radiosonde equipment. In 1947, with New York University, he launched a research balloon called *Mogul,* which landed near Roswell, New Mexico, giving rise to stories that an alien spacecraft had crashed there. *Harris F. Smith Collection*

142 EPILOGUE

Naval officers Malcolm D. Ross and M. Lee Lewis (left to right) followed up Moore's pioneering manned ascent. *Courtesy Mrs. Malcolm D. Ross*

They, too, took off from the Stratobowl. After reaching peak altitude, the 2-million-cubic-foot polyethylene bag began losing altitude. The problem was a faulty helium valve but the Navy fliers didn't know that. They thought the envelope had failed.

As *Strato-Lab I* fell back toward earth, Lewis radioed to the ground that they had an emergency. Capt. Norman L. Barr, a Navy medical officer who was part of the project, monitored transmissions from the balloon as he circled underneath in a Navy plane. He cautioned against trying to land the gondola or the crew by parachute, given the nature of the Bad Lands below. The black and white gondola was carried by the balloon via a sixty-four-foot cargo parachute that could work for Ross and Lewis once they were in lower and denser air. They also had chutes of their own.

The balloon continued to fall, slowing as it reached lower altitudes. At seventeen thousand feet, where gondola internal pressure and outside air pressure were the same, hatches were opened. Everything that could be was thrown overboard—the instrument panel, air regenerator, radio, everything. Gradually the balloon was brought somewhat under control.

Gunpowder squibs were electrically fired, which severed the connections between gas bag and cargo chute, thus freeing the balloon from its load. (Such squibs were introduced to ballooning by Jean Piccard.) Descending eight hundred feet a minute, the gondola hit the ground near Brownlee, Nebraska.

A record had been set but not officially. There was no barograph on board, an essential instrument if the FAI was to "homologate" a new record height.

The feasibility of Strato-Lab thus demonstrated, Ross and Lewis went up again and reached some 85,700 feet on 18 October 1957.

The U.S. Air Force had already been there. Joseph Kittinger Jr. had made it to ninety-six thousand feet on 2 June 1957 and David Simons to 101,500 feet on 19 August 1957. Each had done so solo in a polyethylene balloon. Kittinger test flew and checked out the balloon system. Simons, a U.S. Air Force flight surgeon, made the ascent to gather aeromedical data. Their program was Manhigh, a research effort led by Col. John Paul Stapp.

As it evolved, Strato-Lab would concern itself increasingly with medical and human engineering experiments. Whereas the Air Force was studying how very high altitude flight affected the human body, the Navy was interested in how it affected performance. Ross worked closely with Captain Barr of the Bureau of Medicine and Surgery. Barr, a pilot and flight surgeon, planned the "human side" of the Strato-Lab flights. For example, he had those on board wearing pressure suits to learn how well they could move about and do useful work.

Strato-Lab III, Ross and Lewis, reached eighty-two thousand feet on 26 July 1958. The flight set no record but it was thirty-four hours, longer than any previous Navy or Air Force stratosphere balloon ascent.

The purpose of *Strato-Lab IV* was different. For years, people who peered at the sky with telescopes and instruments wanted to do so from as high up as possible. The earth's blanket of air interfered with their "seeing." So the idea of taking a telescope up in a balloon wasn't new. *Strato-Lab IV* promised to do that—in spades!

144　EPILOGUE

Strato-Labs I, II, and *III* were crewed by Ross and Lewis. Here Ross receives preflight good-luck wishes from Otto Winzen, builder of the balloon system, and Navy Capt. Norman L. Barr, medical adviser to the program. Ross wears plastic around his feet to keep them warm aloft. *U.S. Navy Photo. Courtesy Mrs. Malcolm D. Ross*

Astronomer John Strong of Johns Hopkins University planned a sixteen-inch Schmidt infrared telescope installation atop the Strato-Lab gondola to view Venus and Mars. When Strong was unable to make the flight, Moore, who had been the backup or alternate pilot for the previous Strato-Lab ascents, got his chance to fly. He and Ross took the instrument up to eighty-one thousand feet on 28 November 1959 from the Stratobowl. Difficulties in stabilizing it in flight marred the results.

Meanwhile, Malcolm Ross was busy promoting the stratosphere balloon as a training vehicle for manned space flight. Nowhere on the ground, he claimed, could future astronauts experience what he called "the break-away phenomenon," the sensation of being completely detached from the earth. He had experienced it with Strato-Lab beginning about ten miles up. In a stratosphere balloon, trainees could practice duties and tasks in a realistic (except for the zero gravity) environment. He encouraged all who would listen to use manned balloons for this purpose. He had no takers. NASA had its own ideas about how to train astronauts.

The fifth would be the highest flight in the Strato-Lab series. Whereas the first had been built by General Mills and Winzen Research, *Strato-Lab V* was the product of the latter. Winzen fabricated both the balloon and gondola. The mission was to test a full pressure suit. Ross was pilot, accompanied by Lt. Cdr. Victor A. Prather Jr. of the Naval Medical Research Institute. Each wore the suit. They wore it exposed directly to the pressure, temperature, radiation, and other extremes that would be encountered.

There was no protective, pressurized sphere. The two men rode skyward in what looked like an open cage, six by five by five feet in size. The floor was aluminum sheeting, the sides and frame aluminum tubing. What passed for walls were Venetian blinds, aluminum colored on one side to reflect the heat of the sun and black on the other to absorb it. Ross and Prather could raise, lower, or reverse these blinds to control the temperatures they were experiencing. Their bodies were instrumented with medical monitoring devices.

Strato-Lab V rose at 7:08 AM on 4 May 1961 from the flight deck of the aircraft carrier USS *Antietam,* 130 miles southeast of New Orleans in the Gulf of Mexico. The ship had maneuvered so that it was moving with the wind to create a calm on deck for balloon layout, inflation, and release.

From top to bottom, *Strato-Lab V* measured 480 feet, including plastic-film balloon, orange and white cargo parachute, gondola, and trailing antenna. A relatively small amount of helium was fed into the bag—with altitude it would expand to fill the balloon. The gas collected as a ball inside at the top, giving the craft the appearance of an inverted tear drop.

When launched, the 10-million-cubic-foot giant climbed at one thousand feet per minute. At twenty-six thousand feet, the pressure suits automatically activated.

On the way up, a hissing sound was heard. Was a pressure suit leaking? The sound stopped. Also the rate-of-climb slowed unexpectedly. Was gas escaping from the balloon? There was immense relief, in the air and on the carrier, when the

Strato-Lab V rose from the carrier *Antietam* on 4 May 1961, using the same launch technique as this prior unmanned *Skyhook* from the *Valley Forge*. U.S. Navy. Courtesy Julie Janicke Muhsmann, B.A.G. Corp.

cause was discovered. *Strato-Lab V* was passing through a temperature inversion and the warmer air was reducing its lift. Once through it, the balloon regained its anticipated climb.

Things were going well even when the temperature went to -94° Fahrenheit (at fifty-two thousand feet).

Peak altitude for *Strato-Lab V* was 113,740 feet. It was the highest any manned balloon had ever been. And the highest any balloonist has been since. Its crew looked out in wonder at the vast panorama before their eyes: a half-million square miles of the earth's surface.

The descent that afternoon was uneventful until the very last moments. As they neared the surface of the Gulf of Mexico, Ross fired the squibs that released the balloon. The cargo chute took over and began to lower them the rest of the way. The "cage" splashed into the water. Ross and Prather abandoned it, expecting to be picked up by helicopters from the *Antietam*.

Ross stepped on the rescue hook lowered to him by one of them. His foot slipped. He barely kept from falling by grabbing and hanging onto its cable. The crew hoisted him up and brought him inside.

Prather, too, slipped on the hook of the other helicopter. He did fall and the seawater entered his suit. He drowned. President Kennedy would invite his widow and little boy and girl to the White House to receive Vic Prather's Distinguished Flying Cross, awarded for "heroism and extraordinary achievement."

It was the end of the Strato-Lab program. John Glenn's flight into orbit less than a year later (20 February 1962) made further balloon flights superfluous.

There have been no Navy manned balloons since.

SELECTED BIBLIOGRAPHY

Althoff, William F. *Sky Ships: A History of the Airship in the United States Navy.* New York: Orion, 1980.

———. *USS* Los Angeles: *The Navy's Venerable Airship and Aviation Technology.* Washington, D.C.: Brassey's, 2004.

Crouch, Thomas D. *The Eagle Aloft: 200 Years of the Balloon in America.* Washington, D.C.: Smithsonian, 1983.

DeVorkin, David H. *Race to the Stratosphere: Manned Scientific Ballooning in America.* New York: Springer Verlag, 1989.

Eckener, Hugo. *My Zeppelins.* Translated by Douglas H. Robinson. London: Putnam, 1958.

Grossnick, Roy A. *Kite Balloons to Airships: The Navy's Lighter-than-Air Experience.* Washington, D.C.: Superintendent of Documents, 1987.

Hook, Thom. *Flying Hookers for the* Macon: *The Last Great Rigid Airship Adventure.* Baltimore: Airsho Publishers, 2001.

———. Shenandoah *Saga.* Annapolis, Md.: Air Show Publishers, 1973.

———. *Skyships: The* Akron *Era.* Annapolis, Md.: Air Show Publishers, 1976.

Pace, Kevin, Ronald Montgomery, and Rick Zitarosa. *Naval Air Station, Lakehurst.* Charleston, S.C.: Arcadia, 2003.

Robinson, Douglas H. *Giants in the Sky: A History of the Rigid Airship.* Seattle: University of Washington, 1973.

Robinson, Douglas H., and Charles L. Keller. *Up Ship!: U.S. Navy Rigid Airships 1919–1935.* Annapolis, Md.: Naval Institute Press, 1982.

Rosendahl, Charles E. *Up Ship!* New York: Dodd, Mead, 1931.

Ross, Malcolm D. "We Saw the World from the Edge of Space." *National Geographic Magazine,* November 1961.

Ross, Malcolm D., and M. Lee Lewis. "To 76,000 Feet by Strato-Lab Balloon." *National Geographic Magazine,* February 1957.

Ryan, Craig. *The Pre-Astronauts: Manned Ballooning on the Threshold of Space.* Annapolis, Md.: Naval Institute Press, 1995.

Settle, Thomas G. W. "The Gordon Bennett Race, 1932," U.S. Naval Institute *Proceedings* 59, no. 362 (April 1933): 521–25.

———. "Winning a Balloon Race." U.S. Naval Institute *Proceedings* 55, no. 8 (August 1929): 677–84.

Shock, James R. *American Airship Bases and Facilities.* Edgewater, Fla.: Atlantis Productions, 1996.

———. *U.S. Army Airships, 1908–1942.* Edgewater, Fla.: Atlantis Productions, 2002.

———. *U.S. Navy Airships, 1915–1962.* Edgewater, Fla.: Atlantis Productions, 2001.

Smith, Richard K. *The Airships* Akron *and* Macon: *Flying Aircraft Carriers of the United States Navy.* Annapolis, Md.: U.S. Naval Institute, 1965.

Stehling, Kurt R. *Bags Up! Great Balloon Adventures.* Chicago: Playboy, 1975.

Stehling, Kurt, and William Beller. *Skyhooks.* New York: Doubleday, 1962.

Topping, Dale. *When Giants Roamed the Sky: Karl Arnstein and the Rise of Airships from Zeppelin to Goodyear.* Edited by Eric Brothers. Akron, Ohio: University of Akron, 2001.

Vaeth, J. Gordon. *Blimps and U-Boats: U.S. Navy Airships in the Battle of the Atlantic.* Annapolis, Md.: Naval Institute Press, 1992.

———. Graf Zeppelin: *The Adventures of an Aerial Globetrotter.* New York: Harper and Row, 1958.

———. "When the Race for Space Began." U.S. Naval Institute *Proceedings* 89, no. 8 (August 1963): 68–78.

Van Orman, Ward T. *The Wizard of the Winds.* St. Cloud, Minn.: North Star Press, 1978.

INDEX

Page numbers in *italics* refer to photographs.

Adams, Charles E., 125
Aero Club of America (ACA), 8–9. *See also* National Aeronautic Association (NAA)
Aircraft Development Corporation, 67
air dock, 61
Air Force, 143
Airship Patrol Squadron 12, 117
Airship Patrol Squadron 32, 117
airships, nonrigid (blimps), 1–4, 19, 112, 120, 122–27, *124*, 129–32
airships, rigid, 21, 101, 111
Akron, Ohio, 41, 45, 77, 80, 91–94
Akron, USS, 61, 62, 73, 74, 76–78, 87, 95–96, 112
Altamaha, 125
Alvarez, Luis W., 88
Ambruster, R., 17
Amundsen, Roald, 36
Anacostia, 74–76
Anderson, Orvil A., 94, 139, 140
Andrus, C. C., 14
Antietam, USS, 145, 146, *146*
Army, 5, 6, 21–22, 80, 111, 112
Arnstein, Karl, 72, *73,* 74

BagVets, 5
Baker, Newton D., 22
ballonets, 3
balloons: decline in racing, 39, 105–6; end of program, 147; improvements in racing, 46; at Lakehurst, 112, 119–20; launching of free, *121–23;* manned scientific, 133–47; at Moffett Field, 119–20; piloting, 15–16; popularity of racing, 13; postwar use, 5–9, 128–29; preparation for racing, 14–15; purchase, 1; revival of racing, 45–46, 136. *See also* James Gordon Bennett International Balloon Race; National Balloon Race
Barnaby, Ralph S., 74
Barr, Norman L., 142–43, *144*
Basel, Switzerland, 80, 83, *84*
Bauch, Charles E., 50, 59, 60
B-class ships, 3–4, *4,* 19
Belgica, 9, 16–17, 102–3
Bennett, James Gordon, 7–8. *See also* James Gordon Bennett International Balloon Race
B. F. Goodrich Company, 3, 5
Birmingham Semi-Centennial, 13
Blair, Roland J., 76, 84
Blaugas, 56–57, 60, 78
blimphedrons, 118, 119
blimprons, 118, 119
blimps. *See* airships, nonrigid (blimps)
Blimp Squadron 14 ("the Africa Squadron"), 124
Blimp Squadron 41, 122
Blimp Squadron 42, 122, 124–25
Brackett, Fay. *See* Settle, Fay (Brackett)
Bradley, M. M., 65
Brazil, 85–86, 120, 122, 134
Bureau of Aeronautics (BuAer), 25, 74, 85
Burgess, Charles P., 62, 87
Burzynski (Polish balloonist), 105
Bushnell, Wilfred, 64–65, 69–71, 76, 80–85, *81,* 96, 102
Byrd, Richard E., 35

Campbell, C. I. R., 23
Cape May, N.J., 4, 21
Cavell, Edith, 16
C-class ships, 19–20, *20*
Century of Progress, A, 88–94, *90*
Chatham, Mass., 4, 21
Chicago World's Fair, 87–94

City of Akron, 13
City of Cleveland, 54
Class A (Spherical Balloons), Fifth Category, 65
Coast Guard, 134
Cody, Ernest D., 125
Coil, Emory W., 19, 23
Compton, Arthur H., 88
Connecticut Aircraft Company, 1, 3
Coolidge, Calvin, 32, 59
Cooper, J. F., 54
Copeland, R. W., 50
Crompton, George, 19
Curtiss Aeroplane and Motor Company, 3
Curtiss Marine Trophy Race, 74

Dammann, Carl W., 6
Daniels, Josephus, 2–3, 12
Davis, Robert J., 97
D-class ships, 20
Defender, 111
DeMuyter, Ernest, 9, 16–17, 18, 39, 69, *70,* 102–3
Dennett, R. R., 76
diesel engines, 60
dirigibles. *See* airships, rigid
DN-1, Dirigible, Navy #1, 1–2, *2*
Doenitz, Karl, 127
Dollfus, Charles, 85–86
Douglas Leigh Sky Advertising Company, 128
Dow Chemical Company, 88
drag ropes, 64–65
Dresel, Alger H., 78, 95–98
duralumin, 23–24

Eareckson, William O., 46, 55, 69
Earhart, Amelia, 111
early warning squadron, 131–32
Eckener, Hugo, 56–60, 65–67, 74, 83, 85, 107, 108
Eckener, Knut, 57
E-class ships, 20
Edison, Charles, 113
Elder, J. P., 13
Ellsworth, Lincoln, 36
Emerson, Rolfe, 7
Ent, Uzal G., 54, 64
Enyart, William, *94*

Eppes, M. Henry, 130
Esperia, 18
Evert, Paul, 54
Explorer, 140
Explorer II, 94, 139–40

F9C-2 Sparrowhawks, 98–101, *100*
Farrell, Stephen A., 10–11, 104
F-class ships, 20
Fédération Aéronautique Internationale (FAI), 8, 17, 18, 40, 65, 88, 93
"Fido," 118
Fisher, "Jackie," 2
Flamingo ship, 128
Fleet Airships, Atlantic, 118, 125, 126, 128
Fleet Airships, Pacific, 119, 125
Flemming, Hans, 38
Fordney, Chester L. "Mike," 91–94, *94*
France, 30–31
Fuller, Ben B., 92
Fulton, Garland "Froggy," 29, 62

G-1, 111
gasoline, 57, 60
G-class ships, 20
General Mills, Inc., 133–34
Génève, 18
Germany, 30–32
Gilmer, Francis H., 106
Glenn, John, 147
Glut, Marjan, 85
Glynco (Brunswick), Ga., 115
"goldbeater's skins," 24–25
Goodyear advertising blimp, 112
Goodyear Aircraft, 115, 126
Goodyear III, 39–40
Goodyear IX, 104
Goodyear Tire and Rubber Company, 1, 3, 5, 61–62, 83–85, 115
Goodyear-Zeppelin Corporation, 62, 72–74, *75,* 88
Göring, Hermann, 111
Graf Zeppelin, 56–60, *57,* 65–67, 85–86, 107, 111
grapnels, 16–17
Gray, Hawthorne C., 87
Great Britain, 22, 30–31
Grech, Chris, 101
Greenwald, J. A., 76

Grills, Nelson G., 122–24
G-ships, *116,* 119, 126
gunpowder squibs, 143, 146
Gunther, William J. "Beaver," 135–38, *136*

H-1, 21
Halland, H. E., 13, 14
Halsey, William F., 29
Hampton Roads (Norfolk), Va., 4, 21
Hancock, Joy Bright Little, 25–26, 34, 85
Hancock, Lewis, Jr., 26, 33
Hawley, Alan R., 9
Hearst, William Randolph, 66
Heinen, Anton, 26–28
Helios, 134, *135,* 138, *139*
helium, 17, 19, 25, 32, 60, 106, 110–11
Hensley, William N., Jr., 21–22
Hersey, Henry B., 8
Hill, Edward J. "Eddie," 41, 46, 68
Hindenburg, 60, 106–10, *109*
Hinton, Walter, 10–12, 104
Hitchcock, Tex., 115
Honeywell, Homer E., 7, 17
Hoover, George W., 134, 138, *139*
Hoover, Herbert, 59, 63
Hoover, Mrs. Herbert, 77
Houma, La., 115
Houston, 99
Hoyt, H. W., 6
Hudson's Bay Company, 11
Hughes, Charles Evans, 30
Hughes, Howard, 128
hydrogen, 119
Hynek (Polish balloonist), 105

Ickes, Harold, 110–11
Inter-Allied Aeronautical Commission of Control, 22

J-3, 46–48, *47,* 63, 78, 96
J-4, *47,* 63, 78, 111
James Gordon Bennett International Balloon Race: 1913, 1; 1921, 15, 16; 1923, 17–18; 1924–27, 39–41; 1927, 46, 68; 1928, 55; 1929, 69–71; 1930, 80; 1932, 83–85, *84;* 1933, 102–5; 1934, 105; 1935, 105; establishment, 8–9; final, 106; Lansdowne's involvement in, 29; popularity, 6
Janusz, Anton, 106

K-1, 78, 111
K-2, 112
K-47, 122–24
Kendall, Charles H., 102–4
Kennedy, John F., 147
Kepner, William E., 41, 55, 69, 140
Kershaw, E. L., 13
Key West, Fla., 4, 21
King, Ernest J., 96, 98, 100, 102, 124
kite balloons, 5
Kittinger, Joseph, Jr., 143
Klemperer, Wolfgang, 75
Kloor, Louis A., 10–12, 104
Knox, Cornelius V. S., 62
K-ships, 112–15, *116, 117,* 117–19, 122, 125–26, 128, 129

L-1, 112
L-49, 24
L-72 (zeppelin), 21–22
L-8, 125, *126*
Lahm, Frank Purdy, 8
Lakehurst, N.J., 8, 23–26, 30, 35–42, *37,* 41–44, 107–8, 112–13, 115–16, 118–20
Lange, Karl, 66
Lansdowne, Zachary, 21, 28–33, *31,* 35
Lawrence, John B., 18
Lewis, M. Lee, 139–43, *142*
lighter-than-air program, 21, 26, 119, 124, 132
lightning, 53–55, 64
Lipke, Donald L., 49
Litchfield, Paul W., 39, 62, 74
Little, Charles Gray, 25
Little, Joy Bright. *See* Hancock, Joy Bright Little
Litvinoff, Maxim, 93
Los Angeles, 33, 95; construction, 31–32; decommissioning, 78, 111, 113; design, 56; fly by, 63; gas consumption, 57; glider drop from, 74–75; and *Graf Zeppelin,* 58, 60; Rosendahl on, 48–49; Settle on, 38; standing on nose, 41–44, *44;* tour of East Coast, 77; and water recovery, 96
Los Angeles, Calif., 65–67
L-ships, 112, 113, *116,* 119, *120,* 126, 128
Luftschiffbau-Zeppelin, 21–22, 31, 61, 62

LZ-126. *See Los Angeles*
LZ-130, 111

MacCracken, Allen, 71, 80
Macon, USS, 79, 95, 97; construction, 61, 62, 73; crash, 74, 101; flights, 87, 96–100; Mayer on, 76; naming, 78; at West Coast Lakehurst, 79; Wiley on, 98–101
Magnetic Anomaly Detector (MAD), 118, 124
Maitland, E. M., 23
Manhigh, 143
Mark, Thomas, 11
Mattice, 12
Maxfield, Lewis H., 1, 4, 23
Maybach, Dr., 60
Maybach engines, 56, 77
Mayer, Roland G., 62, 76–77
McAdam, John, 65
McCord, Frank C., 78, 95–96
McCrary, Frank R., 1, 26, 28
McKeesport, Pa., 50
McNamee, Graham, 37
Meyer, Adolphus, 134
Miller, C. F., 76
Millikan, Robert A., 88
Mills, George H., 113, 118–19
Mines Field, 65–67
Missouri Air Reserve Field, 6
Mitchell, William "Billy," 21, 25
Mk 24, 118
Mk I eyeball, 118
Moffett Field, 79, 115–17, 119–20
Moffett, William Adger, 25–26, 29, 31, 59, 61, 63, 65, 78, 85, 87
Montauk Point, N.Y., 4, 21
Monterey Bay Aquarium Research Institute (MBARI), 101
Moore, Charles B., 138–39, 140, 141, 144
Morton, Walter, 41, 46, 53–54, 54
M-ships, 116, 125

National Aeronautic Association (NAA), 9, 65. *See also* Aero Club of America (ACA)
National Balloon Race, 6, 13–15, 45–46, 50–55, 51, 63–65, 76, 80–83, 106

National Broadcasting Company, 88
Naval Airship Training and Experimental Command, 119, 125
Naval Air Station, Akron, 3, 28
Naval Air Station, Anacostia, Washington, D.C., 63
Naval Appropriations Act of Congress (1919), 22
Naval Aviator (Lighter-than Air) designator, 4
New York Times, 58
Nobile, Umberto, 36
Nordman, Rudolf, 40
Norfleet, Joseph P., 17
Norge, 36
North Pole, 35–36

Officers Airship Training School, 136
Ohio Insulation Company, 64
Olmstead, Robert, 17–18
Orville, Howard T., 105
Osoaviakhim, 94
Osten, Arthur, 64
Outlaw, The, 128

Palos, USS, 102
Patoka, USS, 29, 30, 32, 48
PBY Catalina flying boats, 124
Peary, Robert E., 36
Penoyer, Ralph "Horse," 29
Pensacola, Fla., 4, 21
Piccard, Auguste, 88, 140
Piccard, Bertrand, 140
Piccard, Donald, 140
Piccard, Jean, 89, 133, 139, 139–40
Piccard, Jeanette, 139, 139–40
Pierce, Maurice R., 59, 60
Pleiades, 133
Polar, 18
Polish balloonists, 105–6
pontoons, 53, 82
Portland, 119
Post, Augustus, 9
Prather, Victor A., Jr., 145–47
Pratt, Admiral, 83
pressure suit, 145
pressurized cabin, 87
Preston, R. A. D. "Rad," 1, 9

Project Durante, 138
Prüfling, 74, 75
Pruss, Max, 108

R-34 (airship), 21
R-38 (airship), 22–23, 30, 95
radar, 118
radio, 46, 50
Reed, W., 17
Reeves, Joseph N., 100
Reichelderfer, Francis W., 6, 18, 50, *51*, 108
Review Club of St. Louis, 14
Richardson, Jack C., 65
Richmond (Miami), Fla., 115, 128
Rockaway Beach, N.Y., 4, 20
Roma, 112
Roosevelt, Franklin, 99, 100
Roper, 118
Rosendahl, Charles E. "Rosie," *113;* on *Akron*, 78, 95; career, 43–44; on *Graf Zeppelin*, 58, 65; and *Hindenburg*, 108, 110; at inaugural parade, 63; on *Los Angeles*, 48–49; and Naval Airship Training and Experimental Command, 119; preparation for war, 112; on *Shenandoah*, 76; on substitution for Catalinas, 124; on zeppelin design, 74
Ross, Malcolm D., 139–46, *142, 144*
Ross, Margaret Selden, 29
Roth, L. J., 13, 14

Santa Ana, Calif., 115
Saratoga, 48
Savoie, 18
Schlosser, A. G., 68
"Sea Scouts," 2. *See also* airships, nonrigid (blimps)
Seiberling, Willard F., 13, 14
Settle, Fay (Brackett), 36, 86
Settle, T. G. W. "Tex," *70, 73, 94;* on *Akron*, 78; on Atwater Kent radio, 46; at Bureau of Aeronautics's Airship Design Section, 62–63; command of Fleet Airships, Pacific, 119; end of racing career, 105; glider flight, 74–75; in Gordon Bennett races, 69–71, 83–85, 102–5; on *Graf Zeppelin*, 59–60, 66, 67, 85–86; interest in flying, 61, 62; at Lakehurst, 35–38, 42; in National Balloon Races, 45, 50, 52–53, 55, 63–65, 76, 80–83, *81;* personal characteristics, 38–39; saving of J-3, 46–48; on *Shenandoah*, 76–77; stratosphere flights, 87–94
Shenandoah, USS, *27, 30*, 95; accident, 32–33; christening and early flights, 26–28; construction, 23–25, *24;* gas consumption, 57; at Lakehurst, 35–38; Lansdowne as commander, 29–33; Settle on, 76–77; sharing of helium, 32; and water recovery, 96
Shock, James R., 5
Shoptaw, John, 17–18
Simmons, David, 143
Sloman, Frank, 7
Smith, Harris F., 134–38, *136, 139*
South Weymouth, Mass., 115
Soviet Union, 93–94
Stapp, John Paul, 143
Steele, George W., 32
Steelman, George N., 45, 50, 55
Stevens, Albert W., 94, 139, 140
Stevens, J. H., 50, 55
St. Louis, 17
Stratobowl, 140, 142
Strato-Lab, 140–47
Strato-Lab I, 142–43
Strato-Lab III, 143
Strato-Lab IV, 143–44
Strato-Lab V, 145–46, *146*
stratosphere flights, 87–94, 133, 139–46
Strong, John, 144
Sunnyvale facilities, 115–16. *See also* Moffett Field

TC-13, *111*, 111–13, 117
TC-14, 111–12, 117
Thompson, Richard, 7
Tillamook, Oregon, 115
"Tin Blimp." *See* ZMC-2 (Lighter-than-Air, Metalclad)
Tobin, Frederick J., 109–10
Tressey, W. H., 13
Trimble, South, Jr., 110

Trotter, Frank A., 76, 102, 104–5
Tyler, Raymond F., 105, 106

U-134, 122–24
U-843, 124
U-853, 127
U-boats, 1, 122–27, 130–31
Union Carbide and Carbon, 88
United States (balloon), 8
Upson, Ralph H., 1, 6, 7, 9, 13–14, 67–68
U. S. Army Airships, 1908–1942 (Shock), 5
USSR, 94
USSR-lbis, 94

Van Gorder, Harold B., 130
Van Orman, Ward, *54, 70;* and *Century of Progress* flight, 89; end of racing career, 105; in Gordon Bennett races, 17, 39–41, 69, 71, 76, 80, 83, 84, 102, 104–5; on *Graf Zeppelin,* 85–86; and lightning strike, 53–54; in National Balloon Races, 6, 13, 14, 46, 63–65; pontoons, 82; relationship with Settle, 65
variometer, 15
Vaterland, 40
Verheyden, Edward J., 6
Ville de Bruxelles, 17–18
von Hindenburg, President, 60
von Hoffman, Bernard, 14
von Meister, Friedrich Wilhelm von, 59–60, 66–67, 74
von Opel, Fritz, 103

Watson, George F., 44, 50, 55
weather information, 46, 83–84

Weeksville (Elizabeth City), N.C., 115
West Coast Lakehurst, 78, 79
Westover, Oscar, 17
Weyerbacher, Ralph, 26, 28
Whittle, George V., 62
Wick, Zeno, 26
Wiley, Herbert V., 48, 78, 96, 98–101, *99*
Wilkinson, E. W., 13
Williams, Al, 41
Wingfoot Lake. *See* Naval Air Station, Akron
Winzen, Otto C., 133–35, *139,* 139, *144*
Winzen Research, Inc., 139, 145
Wollam, Carl K., 39–40, 54
Wyland, R. V., 13

zeppelin airships, 1, 72–74
Zeppelin Company, 72, 115
Zeppelin, Count, 56
ZMC-2 (Lighter-than-Air, Metalclad), 67–69, *68,* 111
ZNP-K class, 115
ZP5K, 129
ZPG-2s, 129–30, *130*
ZPG-2W, *130,* 131–32
ZPG-3W, *131,* 131–32
ZPN-1, 129
ZR-1. *See Shenandoah,* USS
ZR-2 (Lighter-than-Air, Rigid #2), 22 23
ZR-3 (Lighter-than-Air, Rigid, #3). *See Los Angeles*
ZRS-4 (Lighter-than-Air, Rigid, Scout, #4). *See Akron,* USS
ZRS-5. *See Macon,* USS

ABOUT THE AUTHOR

Lt. J. Gordon Vaeth, USNR (Ret.), served in Navy Lighter-than-Air during World War II as air intelligence officer for Commo. George H. Mills, Commander, Fleet Airships, Atlantic. At the war's end, he was assigned to the staff of Rear Adm. Charles E. Rosendahl, Chief of Naval Airship Training and Experimentation, as historian and officer-in-charge of the new Naval Airship Museum.

In 1946, Rosendahl sent Vaeth to seek out European balloon and airship experts (the German Zeppelin Company's Dr. Hugo Eckener in particular) who had survived the conflict. The task took him to the Congress of Aeronauts in Switzerland where balloonists from Britain, France, the Netherlands, and Switzerland were gathered to plan for the resumption of the prewar Gordon Bennett balloon races.

Being the only American present, Vaeth was quickly pressed into service to promote that goal in the United States. To help him, he was made American representative of the International League of Aeronauts.

After his return and release to inactive duty, he began working with the NAA to revive the Gordon Bennetts. They were never resumed as the world had known them. Gas balloon races called Gordon Bennetts would be held in postwar years but without the fanfare and newsworthiness of the originals. In 1947, Vaeth, a civilian, was invited to join the Office of Naval Research (Special Devices Center) Project *Helios*—a planned manned balloon ascent to one hundred thousand feet—as the flight's scientific research administrator. When *Helios* was terminated, he continued for a while in the unmanned Skyhook balloon program that followed. He became well acquainted with Ross, Lewis, Moore, and Winzen of the Navy's manned balloon program they were getting under way.

Vaeth was also becoming interested in altitudes that balloons could not reach: altitudes in space. In 1958, he became technical staff member for Man-in-Space at the Pentagon's Advanced Research Projects Agency. Later he would become director of systems engineering, also of operations, for the weather satellite programs of the National Oceanic and Atmospheric Administration (NOAA).

In addition to authoring *They Sailed the Skies,* Vaeth has written *Graf Zeppelin* and *Blimps and U-Boats* about lighter-than-air craft.

J. Gordon Vaeth and his wife, Corinne, reside in Olympia, Washington.

The Naval Institute Press is the book-publishing arm of the U.S. Naval Institute, a private, nonprofit, membership society for sea service professionals and others who share an interest in naval and maritime affairs. Established in 1873 at the U.S. Naval Academy in Annapolis, Maryland, where its offices remain today, the Naval Institute has members worldwide.

Members of the Naval Institute support the education programs of the society and receive the influential monthly magazine *Proceedings* and discounts on fine nautical prints and on ship and aircraft photos. They also have access to the transcripts of the Institute's Oral History Program and get discounted admission to any of the Institute-sponsored seminars offered around the country.

The Naval Institute also publishes *Naval History* magazine. This colorful bimonthly is filled with entertaining and thought-provoking articles, first-person reminiscences, and dramatic art and photography. Members receive a discount on *Naval History* subscriptions.

The Naval Institute's book-publishing program, begun in 1898 with basic guides to naval practices, has broadened its scope to include books of more general interest. Now the Naval Institute Press publishes about one hundred titles each year, ranging from how-to books on boating and navigation to battle histories, biographies, ship and aircraft guides, and novels. Institute members receive significant discounts on the Press's more than eight hundred books in print.

Full-time students are eligible for special half-price membership rates. Life memberships are also available.

For a free catalog describing Naval Institute Press books currently available, and for further information about subscribing to *Naval History* magazine or about joining the U.S. Naval Institute, please write to:

Customer Service
U.S. Naval Institute
291 Wood Road
Annapolis, MD 21402-5034
Telephone: (800) 233-8764
Fax: (410) 269-7940
Web address: www.navalinstitute.org